S0-BFD-782

Seven Ideas that Shook the Universe

Seven Ideas that Shook the Universe

Nathan Spielberg
Bryon D. Anderson

Wiley Science Editions

John Wiley & Sons, Inc.
New York • Chichester • Brisbane • Toronto • Singapore

Publisher: Stephen Kippur
Editor: David Sobel
Managing Editor: Andrew B. Hoffer
Composition: The Publisher's Network

Library of Congress Cataloging-in-Publication Data

Spielberg, Nathan.
Seven ideas that shook the universe.

Includes bibliographies and index.
1. Physics. 2. Mechanics 3. Astronomy
I. Anderson, Bryon D. II. Title.
QC21.2.S65 1987 500.2 86-15996

Cloth ISBN 0-471-859-74-5
Paperback ISBN 0-471-848-16-6

Printed in the United States of America

87 88 10 9 8 7 6 5 4 3

Contents

Preface

Many people have heard of such scientific luminaries as Galileo, Newton,and Einstein but do not have clear ideas about what they actually did. For example, in the pop-art world of the television commercial, Einstein, like Charlie Chaplin, helps to sell personal computers; before that he helped to unleash the awesome and sometimes destructive power of the atomic nucleus. We hope that this book will help correct this and similar misconceptions and make clear what really were the great achievements of Einstein and other scientists.

We have adopted a descriptive approach to several major physical concepts that have been developed over the past few centuries, relating them to the historical and philosophical context in which they arose. The emphasis is on the origin, meaning, significance, and limitations of these concepts in terms of our understanding of the physical nature of the universe in which we live. Relatively little is said about applied science and technology except as side or illustrative remarks; nor is there much discussion about detailed implications of science for the benefit or detriment of society. We are not concerned here with the impact of science and technology on society or politics or the environment.

The major concepts dealt with are the Copernican approach to astronomy, the Newtonian approach to mechanics, the idea of energy, the entropy concept, the theory of relativity, quantum theory, and conservation principles and symmetries. Quantitative material and concepts are presented primarily through graphs and diagrams. There are even a few formulas, which are stated but not proved. Sometimes they are justified on the basis of simplicity and plausibility or by analogy.

This book was originally developed for use in college courses satisfying university liberal education requirements. One of the goals of such courses is to introduce students to some of the major developments in science that have affected the culture in which we live. In revising the book for the general reader, we have removed some of the more technical discussions as well as the usual review questions typically placed at the end of each chapter. We have rewritten some of the material and have added some additional discussion on the interconnections of the various scientific advances with thought in philosophy and literature. Nevertheless, our primary focus remains on the physical content of the ideas to be discussed. We believe that one cannot understand the implications or applications of physics to other areas of human thought without a correct understanding of physics. We hope, for example, that readers of this book will understand that Einstein's theory of relativity does not imply that everything is relative, as seems to be the popular impression. We have included in the references given at the end of the book some works interested readers can consult for further discussion of the interconnections (and misconnections) between physics and philosophy, literature, and the arts. These works may also serve as guides to further reading.

As in all books of this type, it is difficult to acknowledge all the colleagues, sources, and books we have consulted in addition to those specifically mentioned in the references or figure captions. We would like particularly to acknowledge the comments and help we have received from Professors David W. Allender, Wilbert N. Hubin, John W. Watson, and Bobby Smith of Kent State University; Carol Donley, Hiram College; Robert Resnick, Rensselaer Polytechnic Institute; Dewey Dykstra, Jr., Boise State University; Ronald A. Brown, SUNY Oswego; Jack A. Soules, Cleveland State University; Nelson M. Duller, Texas A & M University; Thad J. Englert, University of Wyoming; and Eugen Merzbacher, University of North Carolina. We appreciate also the help of David Sobel and Robert McConnin, editors at John Wiley & Sons, and their staffs. Joan Anderson carried out the task of manuscript preparation, and we are grateful to her, Alice B. Spielberg, and William J. Ambrogio for the preparation of an early version of our original book. Of course, our students gave us cause to prepare this material in the first place. As is always true, errors of omission and commission are our responsibility.

N. Spielberg
B. D. Anderson

Horsehead Nebula in Orion
(Photo courtesy Mount Wilson and Palomar Observatories)

1

Introduction

Matter and motion: The continuing scientific revolution

A certain glamour surrounds revolutions and revolutionary ideas. Participants in political revolution often believe that by discarding the established order they are heroically casting off shackles and bravely achieving some new freedom. Scientific revolutions carry with them an intellectual glamour and represent the overthrow of a particular way of understanding the physical world in which we live; the cast-off shackles had somehow interfered with a proper prospective of the material world. Scientific revolutions, perhaps even more than political ones, have profound, long-lasting, often unexpected effects on the way in which we view and deal with our surroundings. In that sense, revolutionary scientific ideas can be looked upon as shaking the intellectual universe. This book offers an introduction to seven of the most important and revolutionary ideas of physics.

Revolutions and Science

Individual participants in the development and elucidation of scientific thought over the past 500 years may well be regarded as among the world's greatest and most successful revolutionaries. The revolutions they brought about did not take place overnight, in many cases requiring decades if not centuries for their completion. Nevertheless, their work has led to new ideas and new conceptions of the world and the universe, and the place of humans therein. Their successes have profoundly influenced modes of thought and reinforced belief in reason and rationality as operative tools for understanding the universe.

Virtually all social and political philosophies embrace or react to specific scientific concepts such as energy, relativism-absolutism, order-disorder, determinism-uncertainty. Science, justifiably or not, has often been used to validate or to discredit various aspects of religious thought, and the "scientific method" is considered by many the way to deal with almost all human problems. This book describes the development of the major physical ideas that have contributed to the modern view of the universe and to the general acceptance of the scientific method.

The oldest and most developed of the sciences—and a model for all the others—is physics. Physics addresses the most fundamental questions regarding the nature of our physical universe. It asks such questions as What is the origin of the universe? How has the physical universe evolved and where is it going? What are the basic building blocks of matter? What are the fundamental forces of nature? Because physics is the study of these and other basic questions, it provides the underpinnings for all of the physical sciences, and indeed for all of the biological sciences as well. The ultimate (that is, microscopic) descriptions of all physical systems are based on the laws of the physical universe, usually referred to as "the laws of physics."

Because physics is the most fundamental science, no scientific conclusion that contradicts physical principles can be accepted as ultimately valid. For example, some of the scientific attacks that initially greeted Charles Darwin's ideas of biological evolution were based on the branch of physics known as thermodynamics. Early calculations of the length of time it would have taken for the Earth to cool from a molten state to its present temperature indicated that there had not been enough time to allow the necessary evolutionary processes to take place (and Darwin was quite perturbed when informed of this). Only with the recognition in the late nineteenth century that radioactivity could supply additional internal heat was it possible to show that the Earth could actually have cooled slowly enough to allow time for evolutionary processes to take place in Darwinian terms. Radioactive dating techniques, derived from nuclear physics, now give additional support to this idea.

Indeed, it has been said that all science consists of two parts: physics and butterfly chasing. This overstatement emphasizes two aspects of scientific endeavor: the *collection* and *classification* of descriptive material and the *understanding* of the reasons for the various phenomena in terms of fundamental concepts. All science, including physics, necessarily includes both aspects. In physics perhaps more than any other science, however, real progress is considered to have been made only when understanding of the phenomena has been achieved.

The great scientific concepts to be discussed in this book have been selected both because of their fundamental nature and because of their inherent appeal. They represent major turning points in the development of physics. These concepts will be presented in broad and introductory terms, in the hope that the reader will acquire some understanding of their central themes, how they developed, and their implications for our understanding of our universe. The reader will also learn something about the terminology of physics. Specific discussions of the scientific method and the philosophy of science will be rather cursory, even though such concerns are important for establishing the validity of scientific concepts. Moreover, in this book, there is relatively little discussion of the practical fruits or products of the scientific enterprise, even though it is these fruits—ranging from video games and hand calculators to the preservation and extension of life (and to the horrors of nuclear warfare)—that motivate very intensive cultivation of physical science throughout the modern world.

Dominant Themes in Physics

Two dominant themes run through the development of physics: (1) matter and motion and (2) the search for order and pattern. The first theme represents the attempt to understand, the second the attempt to classify.

Physics deals with matter and motion. Sometimes greater emphasis is placed on one than on the other and vice versa, but it is also important to study the interplay between the two. Indeed, in the modern theories of relativity and quantum mechanics the distinction between matter and motion is blurred. In its ultimate fine structure, matter is considered to be always in motion. Even in the circumstances in which causative motion is considered to cease, matter is still in a state of motion called zero-point motion or zero-point energy.

Physics does not properly concern itself with such ideas as "mind over matter"—that is, the use of the mind or of a Mind to control matter and motion —but rather with the perception of matter and motion. Physicists sometimes concern themselves with the "true nature of reality," but such considerations are more often left to philosophers. There are limits to scientific understanding, and one of the goals of this volume is to indicate what some of these limits are, at least at this point in the development of science.

As a branch of human knowledge, science is concerned with the classification and categorization of objects and phenomena. It constantly searches for relationships among the objects and phenomena, and makes a constant effort to depict these relationships by means of geometrical representations such as diagrams, graphs, "trees," flowcharts, and so on. Often symmetries and recurrences are observed in these relationships, so that attention becomes focused on the relationships as much as on the objects or phenomena that are the subjects of the relationships. Science always desires to find the simplest and most universal general relationships. This naturally leads to models and theories that simplify, exemplify, and expand various relationships.

For example, it is sometimes said that an atom is like a miniature solar system in which the nucleus plays the part of the Sun and electrons play the part of the planets. In accordance with this solar-system model, one can introduce the possibility that the nucleus and the electrons "spin" in the same way as their counterparts in the solar system. Thus it becomes possible to explore the extent to which the same relationships develop in the atom as in the solar system. The solar system serves as a model of the atom, albeit a flawed model, and thus contributes to our understanding of the atom.

Continuing Evolution of Scientific Knowledge—The Seven Ideas

Scientific knowledge is based on experiments by humans, not given in one awe-inspiring moment. There will always be further experiments that can be carried out to increase the amount of scientific knowledge available. The cumulative result of these experiments may be the perception of new patterns in the store of scientific knowledge or the recognition that patterns which had once been thought to be all-embracing were not so universal after all. Previously established notions may be overthrown or greatly modified as new or different patterns of phenomena are recognized. It is this ongoing turmoil and upset, revision and updating, of scientific conceptions that led to the development of the seven major ideas of physics to be discussed in this book. A brief review of these ideas will indicate their major themes.

I. The Earth is not the center of the universe: Copernican astronomy.

For approximately 2000 years, roughly from the time of Aristotle until somewhat after Columbus' voyages to the New World, it was believed that the Earth was at the center of the universe, both literally and figuratively, in reality and in divine conception. The first of the major scientific upheavals to be discussed involved the revival, establishment, and extension of a contrary notion: The Earth is but a minor planet among many planets orbiting the Sun, which itself is but one star rather remote from the center of a typical (among multitudes of others) galaxy in an enormous, almost infinite (perhaps) universe. This first scientific revolution was possibly the most traumatic, not only because of the reception that greeted it, both within and without the intellectual world, but also because it contained some of the seeds and motivation for other revolutionary ideas to come. Not only did Copernicus revise astronomy but he also made use of ideas of relative motion and of simplicity of scientific theories.

II. The universe is a mechanism that operates according to well-established rules: Newtonian physics.

All objects in the universe are subject to the laws of physics. Isaac Newton, in his presentation of his laws of motion and his law of universal gravitation, succeeded in showing the very solid physical foundation underlying the ideas of Copernican astronomy. He, and those who built upon his ideas, showed further that these and other similar laws underlie the operation of the entire physical universe, both as a whole and in the minutest detail. Furthermore, these laws are comprehensive and comprehensible. As will be apparent from the detailed discussion in Chapter 3, this second idea, which associates cause and effect (causality), aside from its governance both of natural and artificial phenomena, processes and devices, has far-reaching implications for the two contrary doctrines of determinism (predestination) and free will.

III. Energy drives the mechanism: the energy concept.

Despite its all-encompassing nature, Newtonian physics—the science of mechanics—does not provide a completely satisfactory description of the universe; we need to know what keeps the wonderful mechanism running. The ancients considered that Intelligences, or even a Prime Mover, kept everything going. The third idea says that Energy keeps the universe going. Energy appears in various forms, which can be converted into each other. The recurring energy crises that cause so much alarm actually represent shortages of some one of these forms of energy and problems resulting from the need to convert energy from one form to another. We can get some idea of what energy "really" is by comparing it to money. Money is a means of exchange and interaction between humans. So, too, energy is often exchanged in the interactions among various objects in the universe. Just as there is usually a limit on the amount of money available, there is also a limit on the amount of energy available. This limit is expressed as a conservation principle, which governs the dispensation of energy in its various forms. (The principle of conservation of energy is often stated "Energy is neither created nor destroyed but merely changes from one form to another.") As will be discussed in detail in Chapter 8, quantities besides energy are also exchanged in interactions between physical objects and also obey conservation principles.

IV. The mechanism runs in a specific direction: entropy and probability.

Although energy is convertible from one form to another without any loss of energy, limitations on the degree of convertibility nevertheless do exist. A consequence of these limitations on the convertibility of energy from one form to another is the establishment of an overall time-sequential order of past events in the universe. The limitations on convertibility are governed by the same rules that govern the throwing of dice in an "honest" gambling game—the laws of chance (that is, statistics). This suggests the possibility that the revolution inherent in the idea of determinism may be at least modified by a counterrevolution. These considerations indicate that heat, one of the possible forms of energy, must be recognized as a "degraded" form of energy, a conclusion especially important to remember in times of serious energy problems.

V. The facts are relative, but the law is absolute: relativity.

The concepts of the theory of relativity have their roots in some of the arguments that arose during the development of Copernican astronomy. And, although Albert Einstein is now almost universally associated with the theory of relativity, he did not originate the idea that much of what we observe depends upon our point of view (more accurately, our frame of reference). Indeed, Einstein originally developed his theory in order to find those things that are invariant (absolute and unchanging) rather than relative. He was concerned with things that are universal and the same from all points of view. Nevertheless, starting with the then revolutionary idea that the speed of light must be invariant regardless of the reference frame of the observer, he was able to show that many things thought to be invariant or absolute, such as space and time, are relative. Einstein reexamined the basic concepts of space and time and showed that these concepts are closely intertwined. Einstein's work was in fact part of a general reexamination of basic concepts and assumptions going on at that time in both physics and mathematics that showed physicists they could not completely disregard philosophical and metaphysical considerations.

VI. You can't predict or know everything: quantum theory and the limits of causality.

This idea, which rejects the complete and meticulous determinism proclaimed as a result of Newtonian physics, arose innocently enough from the attempt to achieve a sharper picture of the submicroscopic structure of the atom. At the beginning of the twentieth century scientists recognized that atoms are built up from electrons and nuclei and efforts were made to obtain more precise information about the motions of the electrons. The sharper picture is not attainable, however, and it is necessary to consider very carefully just what can be known physically and what is the true nature of reality. Although we cannot obtain an extremely sharp picture of the substructure of the atom, we do have an extremely accurate picture. It is thus necessary to use new ways to describe atoms and nuclei: Systems can exist only in certain "quantum states" and measured quantities are understood only in terms of probabilities. This new "fuzzy" picture makes it possible to have a detailed understanding of chemistry and to produce such marvels as transistors, lasers, microwave ovens, radar communications, superstrong alloys, antibiotics, and so on.

VII. Fundamentally, things never change: conservation principles and symmetries.

Our seventh revolutionary idea, which contradicts the common notion that all things must change, is still developing, and the full range of its implications is not yet clear. It does say, however, that some quantities are "conserved": they remain constant or unchanging. Despite the limitations imposed by the preceding idea, physics continues to probe for the ultimate structure of matter, in which immense amounts of energy are involved. It may be that the elusive ultimate building blocks of nature are now known and that they include particles such as quarks, from which protons and neutrons are made. The question naturally arises as to what rules govern the ultimate structure and what these rules reveal about the nature of the physical universe. As already noted, other conserved quantities besides energy affect interactions among the ultimate constituents of matter, in accord with specific rules or conservation principles. Mathematically, each of these rules represents a certain symmetry. All of the recent progress in determining the basic building blocks of matter has followed directly from the recognition of the intimate relationship between conservation laws and symmetries in our physical universe. Since this idea is still evolving, other questions arise: Is it possible to separate matter and energy from the space-time manifold in which they are observed? What is the nature of space-time, and how does its "shape" and "symmetry" affect the various conservation principles that seem to be universally operative? For that matter, is there really an ultimate structure or basic unifying principle at work in the physical universe? We may never know the ultimate answer, but physicists constantly seek to find order in the chaos that lies beyond the limits of our understanding.

There is no reason to suppose that these seven revolutionary ideas are the only ones which will ever develop for physics. It is possible to compare physics to an unfinished symphony in the sense that new "movements" will appear. These new movements will invoke recurring themes as well as new ones, and themes that will grow out of previous movements. For example, the search for symmetries in the most advanced areas of elementary particle physics is not very different from the search for perfection in the physical world that characterized Greek science.

Because of both recurring and new themes and their interplay, physics can be looked upon as "beautiful" in the way symphonies or works of art can be looked upon as intellectually beautiful. (Frequently enough it is the very simplicity of physical ideas that is considered beautiful.)

There are several ways a symphonic composition can be presented to an audience. Sometimes only one or two movements are played, or the score may be rearranged from its original order. In this book the movements of our symphony, the seven ideas, appear more or less in the order in which they developed during a period of twenty-five hundred to three thousand years. Although from the standpoint of logical development from basic principles it is not necessarily the best order, this chronological presentation does lead to easier understanding of the ideas and better recognition of the context in which they developed. It also permits deeper insight into the continuing evolution of the science of physics.

The historical order also coincides roughly with the decreasing order of size of objects studied (the first were the stars and the planets, the most recent subatomic and subnuclear particles). With the decreasing size of objects under study, paradoxically enough, goes increasing strength of forces acting upon these objects, from the very weak gravitational force at work in the universe at large to the exceedingly strong force that binds together quarks (constituents of nucleons and mesons). And also paradoxically, as increasingly basic and fundamental aspects of matter and energy are studied, the discussion becomes increasingly abstract.

Physics Without Mathematics?

The presentation in this book is nonmathematical. The reader with little training in or taste for mathematics should be able to follow the conceptual development. Yet physics is a very quantitative science, and owes its success to the applicability of mathematics to its subject matter. The success and failure of physical theories depend upon the extent to which they are supported by detailed mathematical calculations, and an intelligent discussion of physical concepts cannot ignore mathematics. It is possible, however, to minimize the formal use of mathematics in discussions of science for the general public, as has been demonstrated repeatedly in such publications as *Scientific American* and *Endeavor.* The basic concepts and methods of physics can be communicated quite successfully with a minimal amount of mathematics. Ultimately, nature appears to follow a few amazingly simple rules. These rules—the laws of physics—can generally be described clearly with little formal mathematics. It is the discovery, extension, and application of these laws that requires considerable mathematical ability.

Sometimes it is possible to avoid the use of much mathematics by presenting quantitative relationships by means of graphs. A well-chosen graph, like a picture, is worth a thousand words (and a well-chosen mathematical formula is worth a thousand graphs). Nevertheless, there are a few occasions when it is necessary to present formulas, but this book contains no derivations, brief or lengthy, of formulas.

Analogies are also useful for presenting mathematical relationships. In our discussion of the concept of energy, we used an analogy between energy and money. Another example already encountered is the analogy between the structure of an atom and the structure of the solar system, with atomic electrons playing the role of planets. The use of analogies makes it possible to draw upon common knowledge in discussing difficult concepts, and analogies and models have in fact played a very important role in the development of physics.

To return to our earlier symphonic analogy, it is generally accepted that a truly cultured individual will have some appreciation of music. Such an individual need not be a musician or even a student of music, but must have some familiarity with and appreciation for the themes, rhythms, and sounds of music. This appreciation and familiarity can be acquired without having touched a musical instrument. So, too, a truly cultured individual has some familiarity with and appreciation for the themes, rhythms, and facts of science. It is not necessary for such a person to carry out calculations, simple or difficult. Of course, appreciation for the fine points and overall grandeur of science is greatly enhanced if one knows something about how calculations are carried out, or even more so if one is a scientist; but so is one's appreciation for music heightened if one is a musician. Nevertheless, music is not, and should not be, limited only to people with instrumental ability; and science is not, and should not be, limited only to those with mathematical ability.

It is possible, although difficult, for a talented person with limited mathematical abilities to become an outstanding physical scientist. Michael Faraday, a celebrated nineteenth-century British scientist, was such a person. A self-educated orphan, Faraday made many contributions to both chemistry and physics. He read a great deal, and with his limited mathematical abilities he found analogies extremely useful. He originated the concept of lines of force to represent electric and magnetic fields, using an analogy with rubber bands. He is generally credited with using this analogy to develop the idea that light is an electromagnetic wave phenomenon. James Clerk Maxwell, who subsequently developed the formal mathematical theory of the electromagnetic nature of light, acknowledged that without Faraday's physical intuition he could not have developed his more rigorous theory.

This book is dedicated to the idea that the intelligent public can grasp and appreciate the significance of major physical concepts without having to be led through a morass of mathematics. Reliance is placed on reading ability, not mathematical ability, and the willingness of the reader to take some things on faith. The object here is not to prove but rather to show and tell.

Science and Other Areas of Human Endeavor—Distinctions and Similarities

It will be useful to close this chapter with a comparison of science and other activities and a brief discussion of the scientific method. There are a number of misconceptions that should be overcome, in order to have a better understanding of what science is about.

Surprisingly enough, there are some similarities between science and magic. Aside from questions as to the fraudulent nature of magic, science and magic are usually regarded as opposites, particularly because the knowledge of magic is necessarily limited to a few initiated practitioners and is at least partially irrational. The knowledge and practice of science, by contrast, are available to all who have sufficient amounts of the human faculty of reasoning. Yet, both applied science and magic are motivated by some similar desires, including the desire to understand and to exert control over nature and to bend it somehow to human will. There is always a desire to explain apparently magical or miraculous phenomena in terms of scientific principles. Some branches of science have been derived at least partially from magic—chemistry, for example, from alchemy. There are even certain laws and principles in magic that have parallels in science.

Voodoo practice, for example, makes use of a law of similarity. A witch doctor makes a doll to represent a person and inserts needles into the doll or smears it with filth in order to cause pain or illness for the person represented. In science or technology, mathematical or actual models are constructed to represent an object under study and "experiments" are done with the model to see what would happen to the object or phenomenon under certain conditions. Of course, it is understood that the model is not the actual object, but valuable information is obtained from its use.

Many people regard both science and religion as avenues to truth. In medieval Europe, at least, science was considered the handmaiden of theology, and many people still wrestle with the problem of integrating their scientific and religious beliefs. Often, major new scientific discoveries or advances are hailed as supplying support for or against a particular theological viewpoint. In such concerns, it is important to bear in mind some necessary characteristics of science. Science and scientific conclusions are always tentative and subject to revisions when new evidence is discovered. Therefore, basing religious beliefs on scientific foundations supplies very shaky support for religion.

The goal of science is to explain phenomena in terms that humans can understand without invoking divine inspiration or interference. Scientific knowledge must be verifiable or testable, preferably in quantitative terms. Miracles, on the other hand, cannot be explained scientifically, otherwise they would not be miracles. In general, miracles are not repeatable, whereas scientific phenomena are. The reason science has been so successful in its endeavors is the fact that it has limited itself to those phenomena or concepts that can be tested repeatedly, if desired. Accepting such limitations, however, means that science cannot serve as a guide to human behavior, morality, and will, other than to specify what physical actions are rationally possible. In addition, science cannot logically address the question of the ultimate reason for the existence of the universe at all, although it is astounding how much of the history and development of the universe has been deduced by science.

Science in general, and physics in particular, was once considered a branch of philosophy known as natural philosophy. Metaphysics is the branch of philosophy that deals with the nature of reality, particularly in a secular rather than a theological sense. Such questions as to whether there are unifying principles in nature, whether reality is material or ideal or dependent only upon sensory perceptions, belong to the field of metaphysics. Even such concepts as space and time, which play a great role in formulations of physics, lie within the province of metaphysics. In this sense, philosophy has much to say to physics. (Although individual philosophers, as philosophers, have contributed little to physics.) The experimental results of physics have had, in turn, much to say to metaphysics.

The results of physics also have had some influence (sometimes based on misinterpretation) on ethics, primarily through their influence on metaphysics. Proponents of the idea of moral relativism, for example, borrowed viewpoints from the theory of relativity, much to the chagrin of Albert Einstein.

It is quite important to distinguish between science and technology. A dictionary definition of technology describes it as the "totality of the means employed to provide objects necessary for human sustenance and comfort." Included in these means is applied science. Science, on the other hand, is a systematized body of knowledge, the details of which can be deduced from a relatively small number of general principles by the application of rational thinking. One may think of toothpaste, television, integrated circuits, computers, medicines, automobiles, airplanes, weapons, clothing, and so on, and the devices and techniques of their production, as products or components of technology. The "products" or components of science, on the other hand, are Newton's laws of motion, the photoelectric effect, the invariance of the speed of light in vacuum, semiconductor physics, and so on, and the techniques for their study.

The distinctions between science and technology are not always sharp or clear. Lasers and their study, for example, can be regarded as belonging to the domain of either science or technology. Although the emphasis in science is on understanding, and in technology on application, understanding and application clearly enhance and facilitate each other. It is for this reason that pragmatically oriented societies are willing to spend large sums of money on scientific endeavors, yet it is the almost impertinent curiosity of the human intellect that distinguishes science.

Almost every book, particularly at an introductory or survey level, dealing with physical, natural, or social science, contains a description of the "scientific method," which is supposed to be the valid technique for discovering or verifying scientific information or theories and for solving all sorts of scientific and human problems. The prescription for the scientific method can be summarized as follows: (1) Obtain the facts or data. (2) Analyze the facts and data in the light of known and applicable principles. (3) Form hypotheses that will explain the facts; these hypotheses must be as consistent with established principles as possible. (4) Use the hypotheses to predict additional facts or consequences, which may be used to test the hypotheses further by enlarging the store of facts in step 1. The sequence of steps is repeated systematically as often as necessary to establish a well-verified hypothesis. The scientific method is open-ended and self-correcting in that hypotheses are continually subject to modification in the light of newly discovered facts. If the hypotheses are correct in every detail, no modification is necessary.

Although this is the model procedure for scientific investigation, it is also true that practitioners of the "art" of scientific investigation may not actually follow such a procedure, even in physics. Often hunch or intuition plays a significant role. It is very important to be able to ask the right (significant) questions at the right time as well as to be able to find the answers to the questions. Sometimes it is necessary to ignore alleged facts, either because they are not really facts or are irrelevant or inconsistent (sometimes with a preconceived notion) or are masking other more important facts or are complicating a situation. For example, it is said that Einstein was once asked what he would have done if the famous Michelson -Morley experiment (Chapter 6) had not established the invariance of the speed of light, as required for his theory of relativity. He would have disregarded such an experimental result, he said, because he had already concluded that the speed of light must be regarded as invariant.

Often dumb luck, sometimes called serendipity, plays a role either in revealing a key piece of information or in revealing a particularly simple solution. Such was the case in the discovery of X-rays by Wilhelm Roentgen. (Interestingly enough, he meticulously followed the standard scientific method in pursuing his investigation of the nature of the X-rays, uncovering many significant features, but ultimately came to an erroneous conclusion.)

Ultimately, however, despite the ambiguities and inconsistencies of individual scientists in their own ways, the demands of the scientific method enumerated above must be satisfied. A scientific theory that does not agree with experiment must eventually be either modified satisfactorily or discarded. So too, every theory, however inspired its creator, must satisfy the prescribed requirements of the scientific method. Experiments must be repeated under a variety of conditions, by a variety of people. Results must be consistent with themselves and other results. Working hypotheses must be substantiated. To do less leads to pseudoscience and fraud.

A thorough study of science yields power, both in a large sense and in a very real, detailed, and practical sense. The resulting power is so great there is concern that it not be used for self-destruction rather than for the blessings it can bestow.

Science also gives a deep insight into and understanding of the workings of nature. There is marvelous symmetry and rationality in the physical universe. The material to be presented in the succeeding chapters of this book will not reveal very much about the power of physics; instead it is hoped that from this material the reader will get an idea of the beauty, simplicity, harmony, and grandeur of some of the basic laws that govern the universe. It is hoped as well that the reader's imagination will be both intrigued and satisfied. In terms of the symphonic analogy once more, it is time to end the overture and to enjoy the first movement of the scientific symphony.

Nicolaus Copernicus
(Photo courtesy Yerkes Observatory, University of Chicago)

2 Copernican Astronomy

The Earth is not the center of the universe

Almost everyone believes that the Earth is a planet which travels in a nearly circular path about the Sun and that the Moon travels in a similar path about the Earth. In fact in the present "space age" we look upon the Moon as a natural satellite of the Earth, a satellite that in its motion is no different from the large number of artificial satellites launched from the Earth since 1958. Travel to the Moon is a demonstrated fact, and space travel to and from distant planets—even galaxies—is commonplace in the popular imagination and on motion picture and television screens. Yet four or five centuries ago, anyone who dared express such ideas was considered irrational if not heretical. Indeed, our senses tell us not that the Earth is moving but that the Sun and all the objects in the sky travel in circular paths about the stationary Earth.

Every twenty-four hours the Sun, Moon, planets, and stars "rise" in the east, traverse the heavens, "set" in the west, and disappear from view for several hours before reappearing again. Until just a few centuries ago, the entire body of knowledge available to Western civilization seemed to point to the view that the Earth is both stationary and centrally located in the universe. In this chapter we review the development of this geocentric (Earth-centered) model of the universe and its overthrow and replacement by our modern concepts, a review that will reveal much about the development and nature of scientific thought.

Early Scientific Stirrings in the Mediterranean Area

The origins of Western science are difficult to trace. For our purposes it is sufficient to mention a few generally recognized sources and motivations. Perhaps foremost among these was the Greek penchant for abstraction and generalization. The Greeks, of course, were greatly influenced by their commercial and military contacts with the Mesopotamian and Egyptian civilizations. These cultures had accumulated large bodies of extremely accurate astronomical data and had developed mathematical techniques to employ these data in commerce, surveying, civil engineering, navigation, and the determination of civil and religious calendars. With a knowledge of the calendar it is possible to determine when to plant crops, the favorable seasons for commerce and war, and when various festivals and rites should be held. But the calendar itself is determined by the configuration of the heavenly bodies. (For example, in the Northern Hemisphere, the Sun is high and northerly in the summer, low and southerly in the winter.) Ancient cultures—and some more recent—found great portents for the future in various "signs" or omens seen in the heavens. (On a more practical level, sailors out of sight of land are able to navigate by measurements of the location of various known heavenly objects.)

In general, the ancient civilizations did not make the sharp distinctions between secular and religious affairs common in modern Western society, so they naturally found connections between all aspects of human endeavor and knowledge; mythology and religion; and astronomy, astrology, and cosmology. For Western civilization, the development of ethical monotheism by the ancient Israelites, coupled with the Greek search for a rational basis for human behavior, probably helped motivate philosophers to attempt to unify all branches of knowledge.

The Greek philosophers Socrates, Plato, and Aristotle stressed that civilizations and nations needed to be governed wisely and according to the highest moral principles. This required an understanding and knowledge of the Good. A necessary prerequisite for understanding of the Good was an understanding of science —arithmetic, geometry, astronomy, and solid geometry. It was not enough to be skilled in these subjects; it was also necessary to understand their essential nature as related to the Good. This understanding could only be achieved after long and arduous study. Plato, for example, felt that one could not comprehend the essential nature of such subjects without a thorough mastery of their details and their uses.

According to tradition, the recognition of mathematics as a subject worthy of study in its own right, regardless of its utility for practical matters, was achieved some twenty-six hundred years ago by Thales, who lived on the Asiatic coast of the Aegean Sea. Some time later the followers of Pythagoras (who lived in one of the Greek colonies in Italy) proclaimed that the entire universe was governed by numbers. The numbers they referred to were integers (whole numbers) or ratios of integers. They considered that all things were built up from individual building blocks, which they called *atoms*. Because atoms are distinct units, they are countable—which meant to the Pythagoreans that geometry could be looked upon as a branch of arithmetic.

It was soon realized, however, that there are numbers, such as π and $\sqrt{2}$ that could not be expressed as the ratio of integers. These numbers were therefore called *irrational* numbers, but this caused quite a problem because it meant that many triangles could not be built up of atoms. An isosceles right triangle, for example, might have an integer number of atoms along each of its sides. Its hypotenuse would have $\sqrt{2}$ as many atoms, but this is impossible because $\sqrt{2}$ is irrational and any integer multiplied by $\sqrt{2}$ is also irrational and therefore cannot represent a whole number of atoms. According to legend, the Pythagoreans found the existence of irrational numbers rather disturbing and attempted to suppress the knowledge of their existence.*

It was nevertheless possible to establish connections between arithmetic and geometry through such formulas as the Pythagorean theorem relating the three sides of a right triangle ($A^2 + B^2 = C^2$), or the relationship between the circumference and radius of a circle ($C = 2\pi r$), or the relationship between area and radius of a circle ($A = \pi r^2$), and so on. Some numbers could be arranged in geometric patterns, and relationships were discovered between these patterns. This is shown in Figure 2-1, in which circles are used to represent numbers. Figure 2-la shows the "triangular" numbers such as 1, 3, 6, 10. Figure 2-lb shows the "square" numbers 4, 9, 16. The combination of any two successive triangular numbers is a square number, as shown in Figure 2-lc. The Pythagoreans were said to believe so strongly in the significance of mathematics that they established a religious cult based on numbers.

The assignment of significance to numbers was not limited to the Greeks. According to the Jewish Cabala, the deeper meaning of some words can be found by adding up the numbers associated with their constituent letters: the resultant sum would be significant of a particular subtle connotation of the word. Nowadays, there are "lucky" and "unlucky" numbers, such as 7 or 13. Card games and gambling games are based on numbers. Some buildings do not have a thirteenth floor, even though they may have a twelfth and a fourteenth. In nuclear physics, reference is made to magic numbers, in that nuclei having certain numbers of protons or neutrons are particularly stable and their numbers deemed "magic." In

*If $\sqrt{2}$ were exactly equal to 1.4, then $\sqrt{2}$ would equal the ratio of two whole numbers, 14 and 10—that is, 14/10. If so, then an isosceles right triangle with ten atoms on two sides would have exactly fourteen atoms on its hypotenuse. But $\sqrt{2}$ is actually slightly greater than 14/10, and so the triangle should have more than fouteen (but less than fifteen) atoms on the hypotenuse. It is impossible to find two whole numbers the ratio of which is exactly equal to $\sqrt{2}$. Of course, this problem disappears if the assumption that atoms are the basis of geometry is discarded. This may be one reason that the Greeks never developed the concept of atoms significantly. The concept lay essentially dormant and undeveloped for about two millenia.

atomic physics at one time some effort was expended trying to find significance in the fact that the so-called fine structure constant seemed to have a value exactly equal to 1/137.

The ancient Greeks were also fascinated by the shapes of various regular figures and developed a hierarchy for ranking these shapes. For example, a square may be considered to show a higher order of perfection than an equilateral triangle. If a square is rotated by 90 degrees about its center, its appearance is unchanged (see Figure 2-2). An equilateral triangle, on the other hand, must be rotated by 120 degrees before it returns to its original appearance. A hexagon needs only to be rotated by 60 degrees to preserve its appearance. An octagon needs only a 45-degree rotation, a dodecagon (twelve-sided figure) only a 30-degree rotation.

The more sides a regular figure has, the smaller the amount by which it must be rotated in order to restore its original appearance. In this sense, increasing the number of sides of a regular figure increases its perfection. As the number of sides of such a figure increases, it comes closer and closer in appearance to a circle, and so it is natural to look upon a circle as being the most perfect flat (or two-dimensional) figure that can be drawn. No matter how much or how little a perfect circle is rotated about its center, it maintains its original appearance: its appearance is constant. It is important to note that perfection is identified with constancy—something that is perfect cannot be improved, so it must remain

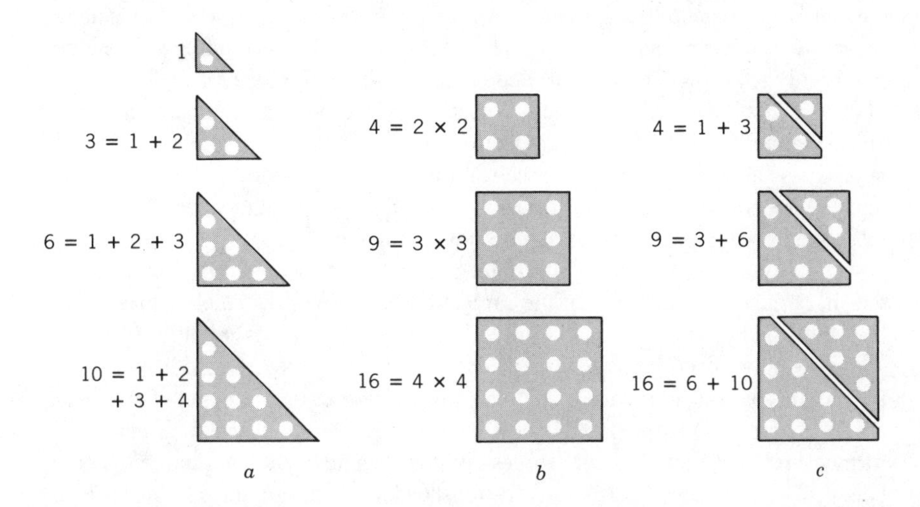

Figure 2-1. Triangular and square numbers. (a) Triangular numbers as arrangements of individual elements. (b) Square numbers as arrangements of individual elements. (c) Square numbers as combinations of two successive triangular numbers in which the first triangle is turned over and fitted against the second triangle.

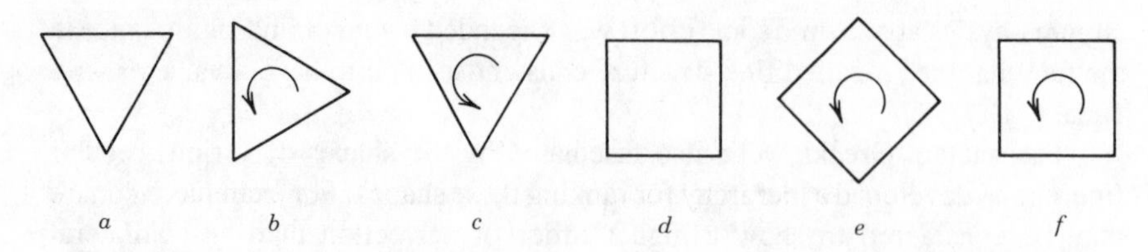

Figure 2-2. Geometrical shapes and symmetry. (a) Equilateral triangle. (b) The triangle rotated by 90°. (c) Rotation of the triangle by 120° from its original position to a new position that is indistinguishable from its original position. (d) Square. (e) Rotation of the square by 45°. (f) Rotation of the square by 90° from its original position to a new position indistinguishable from the original position.

constant. E. A. Abbott's charming little book *Flatland* describes a fictional two-dimensional world peopled by two-dimensional figures whose characters are determined by their shape. Women, for example, the deadlier of the species, are straight lines, and are therefore able to inflict mortal wounds, just as a thin sharp rapier can. The wisest male, in accordance with the Greek ideal of perfection is, of course, a perfect circle.

The Greeks' concern with perfection and geometry set the tone for their approach to science. This is particularly illustrated in Plato's famous Allegory of the Cave. In this allegory, Plato (427–347 B.C.) envisioned humans as being like slaves chained together in a deep and dark cave that is dimly illuminated by a fire burning some distance behind and above them. There is a wall both behind and in front of them. Their chains and shackles are such that they cannot turn around to see what is going on behind them. People moving back and forth on the other side of the wall behind them hold various objects over their heads and make sounds and noises unintelligible to the slaves. The slaves can see only the shadows of these objects, as cast by the firelight on the wall in front of them, and can hear only the muffled sounds reflected from the wall in front of them. The slaves have seen these shadows all their lives, and thus know nothing else.

One day, one of the slaves (who later becomes a philosopher) is released from the shackles and brought out of the cave into the real world, which is bright and beautiful, with green grass, trees, blue sky, and so on. He is initially unable to comprehend what he sees because his eyes are unaccustomed to the bright light. Indeed, it is painful at first to contemplate the real world, but ultimately the former slave becomes accustomed to this new freedom. He has no desire to return to his former wretched state, but duty forces him to return to try to enlighten his fellow beings. This is no easy task, because he must become accustomed to the dark again and must explain the shadows in terms of things the other slaves have never seen. They reject his efforts, and those of anyone else like him, threatening such people with death if they persist (as indeed was decreed for Socrates, Plato's teacher). Plato insisted, however, that duty requires that the philosopher persist, despite any threats or consequences.

The first task of the philosopher is to determine the Reality, or the Truth, behind the way things appear. Plato pointed as an example to the appearance of the heavens, which is the subject matter of astronomy. The Sun rises and sets daily, as does the Moon; and the Moon goes through phases on approximately a monthly schedule. The Sun is higher in the summer than in the winter; morning stars and evening stars appear and disappear. These are the "appearances" of the heavens, but the philosopher (nowadays the scientist) must discover the true reality underlying these appearances—what is it that accounts for the "courses of the stars in the sky"? According to Plato, the true reality must be perfect or ideal, and the philosopher must look to mathematics, in particular to geometry, to find the true reality of astronomy.

Most objects in the sky, such as the stars, appear to move in circular paths about the Earth as a center; it is tempting to conclude that the truly essential nature of the motion of heavenly objects must be circular, because the circle is the perfect geometrical figure. It does not matter whether all heavenly objects seem to move in circular paths or not; after all, human beings can perceive only the shadows of the true reality. The task of the philosopher (or scientist) is to show how the truly perfect nature of heavenly motion is distorted by human perceptions. Plato set forth as the task for astronomy the discovery of the way in which the motions of heavenly objects could be described in terms of circular motion. This task was called "saving the appearances." The goal of discovering the true reality is still one of the major goals of science, even though the definition of true reality now differs somewhat from Plato's definition. This goal has strongly affected the way in which scientific problems are approached. In some cases it has led to major insights; in others, when too literally sought, it has been a severe handicap to the progress of science.

Geocentric Theory of the Universe

If a time exposure photograph of the night sky is taken over a period of a few hours, the result will be as shown in Figure 2-3. During the time the camera shutter is open, the positions of the various stars in the sky change, the paths of their apparent motions being traced by arclike streaks just as photographs of a city street at night show streaks of light from the headlights of passing automobiles. If the camera shutter is left open for twenty-four hours (and if somehow the Sun could be "turned off"), many of the streaks, in particular those near the pole star, would become complete circles. Because the Sun cannot be turned off, the picture can only be taken during the hours of darkness, and thus only a fraction of the complete circles can be obtained, equal to the fraction of twenty-four hours during which the camera shutter is open. If the photograph is taken night after night, it will be almost the same, with a few exceptions to be discussed below.

Figure 2-3. Time exposure photograph of night sky (Fritz Goro/Life Magazine© Time, Inc.)

It appears that the Earth is surrounded by a giant spherical canopy or dome, called the stellar sphere or the celestial sphere, and the stars are little pinpoints of light mounted on this dome (Figure 2-4a). The dome rotates about us once in twenty-four hours in a direction from east to west (rising to setting). The Sun, Moon, and planets are also mounted on this sphere. Of course, when the Sun is in view we cannot see the stars because the bright light of the Sun, as scattered by the Earth's atmosphere, makes it impossible to see the relatively weak starlight. (During an eclipse of the Sun, it is possible to see the stars quite well.)

By patient observation and measurement, the ancient astronomers and astrologers, who did not have photographic equipment, were able to observe and keep track of this daily rotation, called the diurnal rotation or diurnal motion of the celestial sphere. They were able to determine that all the heavenly objects, with few exceptions, are fixed in position on the rotating celestial sphere. In the course of time it was recognized that the Earth itself is also spherical and appears to be located at the center of the celestial sphere. The part of the celestial sphere that can be seen depends upon the observation point on the Earth. At the North Pole of the Earth, the centers of the circles will be directly overhead. At a latitude of 45 degrees the celestial pole (the centers of the circles) will be 45 degrees up from the north horizon (in the Northern Hemisphere), whereas at the Equator the celestial pole will be down at the north horizon.

Fairly early it was recognized that a certain few objects on the celestial sphere were not fixed; these objects appeared to move in relation to the general background stars. The position of the Sun on the sphere, for example, changes over the course of a year. It follows a path, called the ecliptic, shown in Figure 2-4a as a dashed circle tilted at an angle of 23-1/2 degrees from the celestial equator. The

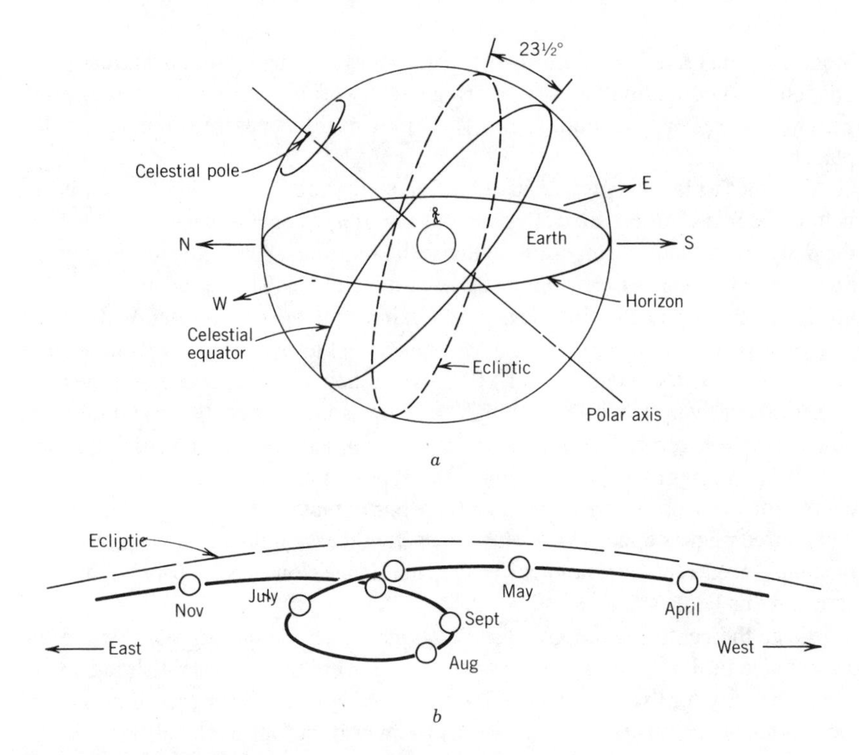

Figure 2-4. Celestial sphere. (a) Schematic of sphere showing Earth at center and the ecliptic. (b) Greatly enlarged view of part of the path of a planet along the ecliptic, showing retrograde motion.

direction of the motion of the Sun is from west to east (opposite the daily rotation).* The position of the Moon also changes, and it travels from west to east along the ecliptic once in 27-1/3 days on the average. Similarly, the planets that are visible to the unaided eye—Mercury, Venus, Mars, Jupiter, and Saturn—travel roughly along the ecliptic, from west to east also, taking ninety days to thirty years to make the complete circuit.

The motion of the Sun is called annual motion, and is fairly, although not completely, smooth and even. The motion of the Moon is also somewhat smooth, but less so than the Sun's motion. The motion of the planets, on the other hand, is variable in speed and sometimes even in direction as well. Indeed, this is the origin of the word *planet*, which means wanderer. When a planet occasionally varies its direction of motion, it seems to move from east to west relative to the fixed stars rather than follow its overall west-to-east path. This motion, called retrograde

*This does not contradict the common observation that the Sun travels from East to West. It simply means that sunrise is always a little late compared to "star-rise" because the position of the Sun on the celestial sphere has changed.

motion, is illustrated in Figure 2-4b, which shows the path of a planet that is not only changing direction but is wandering slightly off the ecliptic as well. Accompanying the retrograde motions are variations in the apparent brightness of the planet.

At times a planet is just ahead of the Sun on the ecliptic; at other times it is just behind the Sun. This is called alternate motion. If the planet is just west of the Sun, the daily rotation will bring it into view as the morning star; if it is just east of the Sun, it will be seen as the evening star just after sunset. (The Moon and Sun were also considered planets, but they do not exhibit retrograde motion or the same degree of variation of speed as the other planets. In this book, the word *planet* does not include the Moon and the Sun.) Because of their varying speeds, at times two planets appear very close to each other, an event sometimes called a conjunction. Ancient astrologers and soothsayers who searched for signs in the heavens attached great significance to conjunctions as portending or reflecting momentous events for humans. Even more rare (and portentous) are double conjunctions, when three planets appear very close to each other. (Conjunction has a different meaning in current astronomical usage, the discussion of which is beyond the scope of this book.)

Over the centuries the great civilizations attached much importance to the appearance of the heavens. These appearances were used to establish the calendars necessary for the governance of extensive empires and for recording important events in human affairs. They were also important for travel to distant places, because navigators could determine where they were by the appearance of the heavens in their particular location. In fact, the calculation and use of accurate navigation tables played a very important part in the exploration of the world by various seafaring nations and in their ability to carry out commercial ventures and military expeditions. Until the last few centuries, astrological predictions, based on the appearance of the heavens in general or some unusual event such as a conjunction or a comet or a nova, have had widespread influence on political decision-making, and horoscopes remain surprisingly popular.

The Babylonians developed one of the great Mesopotamian civilizations. Their astronomers carried out quite accurate measurements and calculations of the heavenly appearances. They were more concerned with the accuracy and reliability of their astronomical measurements and predictions than with the development of a general understanding of the whys and wherefores of the workings of the universe.

The Greeks, although interested in accurate and reliable knowledge of the heavens, did not improve the Babylonian measurements significantly. The Greek contribution to astronomy came more from their concern with understanding on a philosophical level. They regarded the heavens as the regions where perfection could be found, because the true nature of the heavens was perfection. The appearances of the heavens, they felt, had to be interpreted in terms of their inherent perfection and constancy. The Sun and Moon seem to have the "perfect" shape—circular. The fact that the diurnal motion of the stars resulted in circular motion was appropriate because a circle represented the perfect figure, as we have seen.

Although the detailed motion of the planets seemed to depart from smooth and circular motion, on the average the motion was circular. Therefore it was asserted that the essential nature of their motion was circular, and the departures from circular motion that were observed represented but the shadows of the true reality, as understood from Plato's Allegory of the Cave. The Greek philosophers took seriously Plato's command to "save the appearances," to explain all motions of heavenly bodies in terms of circular motions. In modern terms, they undertook to develop a model of the universe which would explain how it "really works."

Many ancient civilizations developed models of the universe. The Egyptians believed the universe to be like a long rectangular box, with the Earth a slightly concave floor at the bottom of the box and the sky a slightly arched iron ceiling from which lamps were hung. There were four moutain peaks supporting the ceiling, and they were connected by ranges of mountains, behind which flowed a great river. The Sun was a god who traveled in a boat along this river and, of course, could only been seen during the daylight hours.

The models considered by the Greeks at a much later time were a great deal more advanced than this and reflected considerably more knowledge. They considered two types of models, the geocentric or Earth-centered type and the heliocentric or Sun-centered type. In the geocentric models, the Earth (in addition to being at the center of the universe) was usually stationary. In the heliocentric models, the Earth circled about the Sun or about a central fire, like all the other planets, and usually rotated on its own axis. By and large, the Greeks preferred the geocentric models and rejected the heliocentric models.

In one of the simplest geocentric models, the Earth is a motionless small sphere surrounded by eight other rotating concentric spheres that carry the Moon, the Sun, Venus, Mercury, Mars, Jupiter, Saturn, and the fixed stars, respectively. Because the spheres all have the same center, this is called a homocentric model. Associated with each of these eight spheres are a number of other auxiliary spheres. (This model is credited to Eudoxus, one of Plato's pupils who took up the challenge to determine the true reality.) The Moon and Sun each have two auxiliary spheres, and the planets have three auxiliary spheres each. Counting the stellar sphere, this made a total of twenty-seven spheres.

The purpose of the auxiliary spheres is to help generate the observed motions. For example, the sphere carrying the Moon rotates about an axis whose ends are mounted on the next larger sphere, so that the axis of rotation of the Moon's sphere can itself rotate with the auxiliary sphere. In turn, the axis of this auxiliary sphere is mounted on the next auxiliary sphere, which can also rotate. By placing these axes at various angles with each other and adjusting the rotation speeds of the spheres, it is possible to generate a motion for a heavenly object so an observer on Earth will perceive that object, against the background of the fixed stars, following the appropriate path along the ecliptic.

In the early versions of this model the question as to how these spheres obtained their motion was ignored. It was simply assumed that the very nature of a perfect sphere in the heavens is such that it would rotate. The purpose of the model was simply to explain the nonuniform motion of heavenly objects as combinations of uniform circular motion, thereby "saving the appearances." To the extent that anyone might wonder about the causes of the motion, it was said to be due to various "Intelligences." By this time the Greek philosophers had more or less put aside such mythical explanations as gods driving fiery chariots across the sky, and so on.

The great philosopher Aristotle (384–322 B.C.) adopted the homocentric model of the universe and thoroughly integrated it into his philosophical system, showing its relationship to physics and metaphysics. He considered the universe spherical and divided into two "worlds," the astronomical (or heavenly) and the sublunar world. The sublunar world was made up of four prime substances—earth, water, air, and fire—whereas the astronomical world contained a fifth substance, ether. These substances had certain inherent characteristics. The laws of physics are determined by the inherent nature of these substances, and therefore the laws of physics in the sublunar world differ from the laws in the astronomical world. Aristotle's understanding of physics was quite different from our understanding of physics today, but it was tightly interlocked with his entire philosophical system. (Aristotle's concepts of physics will be discussed in more detail in the next chapter.)

Aristotle was not content with the simple description of planetary motions introduced by Eudoxus, because in Eudoxus' scheme the motions of the various sets of spheres were independent of each other. Aristotle felt a need to link the motions of the heavenly objects together into one comprehensive system, so he introduced additional spheres between the sets assigned to specific planets. For example, the outermost auxiliary sphere of the set belonging to Saturn had its axis mounted on the sphere of the fixed stars. On the innermost sphere belonging to Saturn was mounted a set of additional spheres, called counteracting spheres, consisting of three spheres that rotated in such a way as to cancel out the fine details of Saturn's motion. On the innermost of these counteracting spheres was mounted the outermost auxiliary sphere belonging to Jupiter. Similarly, groups of counteracting spheres were introduced between Jupiter and Mars, and so on.

All told, in Aristotle's scheme there were fifty-six spheres. The outermost sphere was called the Prime Mover, because all the other spheres were linked to it and derived their motion from it. Aristotle did not specify in detail how the motion was actually transmitted from one sphere to another. Later writers speculated that somehow each sphere would drag along the next inner sphere with some slippage. Many centuries later working models were built to illustrate this motion, with gears introduced as necessary to transmit the motions.

It was soon recognized that there were some weaknesses in the simple geocentric theory discussed by Aristotle. The measured size of the Moon, for example, may vary by about eight or ten percent at different times of observation. Similarly, the planets vary in brightness, being particularly brighter when undergoing retrograde motion. These phenomena imply that the distance from the Earth to the planet or to the Moon changes, but this is not possible in the homocentric system. Even more important, as a result of Alexander the Great's conquests, the Greeks became particularly aware of the large amount of astronomical data that had been accumulated by the Babylonians. They found that Aristotle's model did not fit these data. It thus became necessary to modify Aristotle's model, which was done over the course of several hundred years.

These modifications culminated some 1800 years ago about the year 150, when the Hellenistic astronomer Ptolemy in Alexandria, Egypt, published an extensive treatise, *The Great Syntaxis*, on the calculations of the motions of the Sun, Moon, and planets. With the decline of secular learning accompanying the collapse of the Roman Empire, this work was temporarily lost by the Western world. It was translated into Arabic, however, and eventually reintroduced into Europe under the title of *Almagest* ("the Majestic" or "the Great").

Ptolemy abandoned Aristotle's attempt to link all the motions of the Sun, Moon, and planets to each other, dismissing physics as too speculative. He felt it sufficient to develop the mathematical schemes to calculate accurately the celestial motions without being concerned about the reasons. The only criterion for judging the quality or validity of a particular scheme of calculation was that it should give correct results and "save the appearances." It was necessary to do this because the available data indicated that the appearances of the heavens were gradually changing. To be able to establish calendars and navigational tables, it was important to have accurate calculations, regardless of how the calculations were justified.

To this end, Ptolemy made use of several devices (as the modifications were called) that had been suggested by other astronomers as well as some of his own devising. Nevertheless, he preserved the concept of circular motion. Figure 2-5a shows one such device, the eccentric. This simply means that the sphere carrying the planet is no longer centered at the center of the Earth—it is a bit off center—therefore in part of its motion the planet is closer to Earth than in the rest. When the planet is closer to the Earth, it will seem to be moving faster. The center of the sphere is called the eccentric. In some of his calculations Ptolemy even had the eccentric itself moving, although slowly.

Figure 2-5b shows an epicycle, which is a sphere whose center moves around or is carried by another sphere, called the deferent. The planet itself is carried by the epicycle. The epicycle rotates about its center, whereas the center is carried

Claudius Ptolemy
(The Bettmann Archive)

along by the deferent. The center of the deferent itself might be at the center of the Earth—or, more likely, it too is eccentric. Depending upon the relative sizes and speeds of the epicycle and deferent motion, almost any kind of path can be traced out by the planet. As shown in Figure 2-5c, retrograde motion can be generated, and the planet will be closer to the Earth while undergoing retrograde motion. Combinations of epicycle and deferent motions can be used to generate an eccentric circle or an oval approximating an ellipse, or even an almost rectangular path. (One can visualize epicycle motion as like that of certain amusement-park rides, which have their seats mounted on a rotating platform that is itself mounted at the end of a long rotating arm. The other end of the arm is also pivoted and can move; see Figure 2-6.)

For astronomical purposes, however, it is not enough to generate a particular shape of path. It is also necessary that the heavenly object travel along the path at the proper speed; that is, the calculation must show that the planet gets to a point in

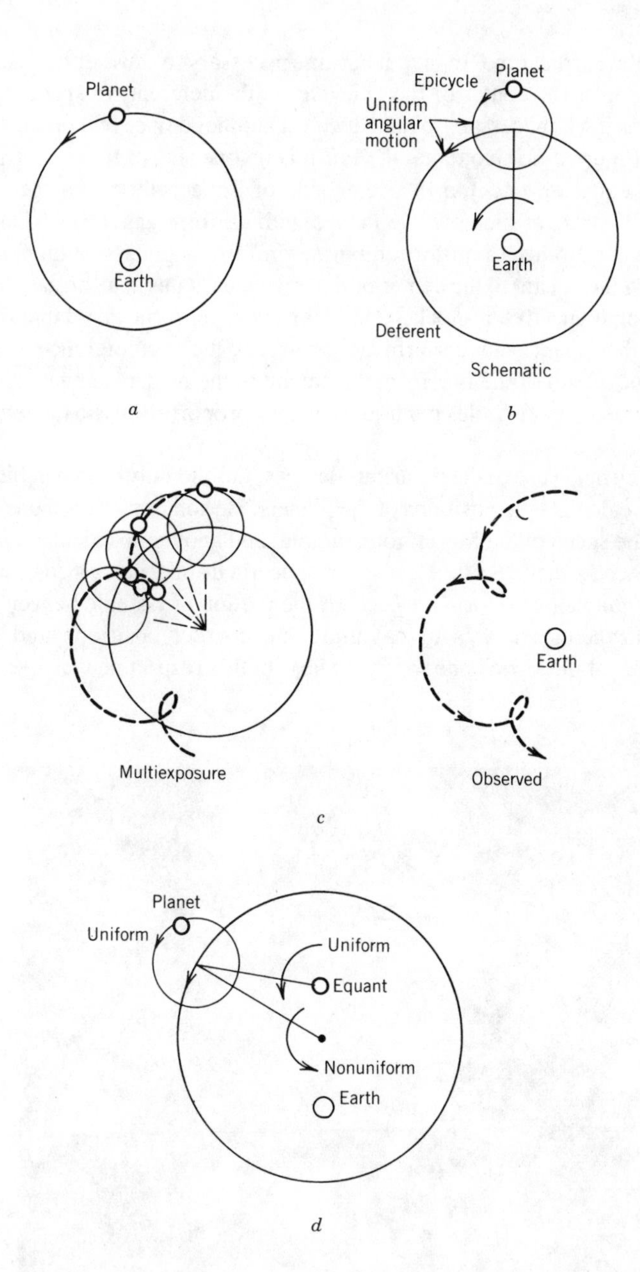

Figure 2-5. Devices used in Ptolemy's geocentric model. (a) Eccentric. (b) Schematic of epicycle on deferent. (c) Generation of retrograde motion by epicycle. (d) Equant.

the sky at the correct time. In fact, it became necessary to cause either the planet on the epicycle, or the center of the epicycle on the deferent, to speed up or slow down. Thus it was necessary to introduce yet another device, called the equant, as shown in Figure 2-5d. Note that the Earth is on one side of the eccentric and the equant an equal distance on the other side of the eccentric. As seen from the equant, the center of the epicycle moves with uniform angular motion, thereby saving the appearance of uniform motion. Uniform angular motion means that a line joining the equant to the center of the epicycle rotates at a constant number of degrees per hour, like a clock hand. However, uniform angular motion with respect to the equant is not uniform motion around the circumference of the circle. Because the length of the line from the equant to the planet changes, the circumferential speed (say, in miles per hour) changes proportionally to the length of the line.

Ptolemy made use of these various devices, individually or in combination, as needed, to calculate the positions of the planets. He sometimes used one device to calculate the speed of the Moon, for example, and another to calculate the change of its distance from the Earth. He was not concerned about using a consistent set of devices simultaneously for all aspects of the motion of a single heavenly object. His primary concern was to calculate the correct positions and times of appearances of the various heavenly bodies. In this respect he was like a student

Figure 2-6. Amusement park ride involving epicycles (Charles Gupton/Stock Boston)

who knows the answer to a problem and therefore searches for a formula that will give him the correct answer without worrying about whether it makes any sense to use that formula.

Some later astronomers and commentators on Ptolemy's system did try to make the calculations agree with some kind of reasonable physical system. Thus they imagined that the epicycles actually rolled around their deferents. This meant, of course, that the various spheres had to be transparent so that it would be possible to see the planets from the Earth. Thus the spheres had to be made of some crystal-like material, perhaps a thickened ether. The space between the spheres was also thought to be filled with ether because it was believed that there was no such thing as a vacuum or empty space.

To the modern mind, Ptolemy's system seems quite far-fetched; however, in his time and for fourteen hundred years afterward, it was the only system that had been worked out and was capable of producing astronomical tables of the required accuracy for calendar-making and navigation. To repeat, Ptolemy himself had one primary goal in mind: the development of a system and mathematical technique that would permit accurate calculations. Philosophical considerations (and physics was considered a branch of philosophy) were irrelevant for his purposes. In short, Ptolemy's scheme was widely used for the very pragmatic and convincing reason that it worked!

The Heliocentric Theory—Revival by Copernicus

According to the heliocentric (Sun-centered) theory, the Sun is the center of planetary motion and the Earth is a planet, like Mars or Venus, that travels in a path around the Sun and also spins on its axis. As early as the time of Aristotle, and perhaps even earlier, the possibility that the Earth rotates on its axis had been considered. There was also an even older suggestion that the Earth, as well as the Sun, moves about a "central fire." These ideas were rejected, and some of the arguments against them are discussed below. The Greeks by and large thought the idea of a moving and spinning Earth was untenable; it was considered impious and dangerous as well as ridiculous to entertain such ideas.

Nevertheless, in the latter part of the fourteenth century, a few individuals suggested that there were some logical difficulties with the concept of the Earth-centered universe. For one thing, if the daily rotation of the stars was due to the rotation of the celestial sphere, because of its huge size the many stars on the surface of the celestial sphere would be moving at impossibly high speeds. It seemed just as reasonable to assume that only the small Earth was spinning and the daily rotation was simply an optical illusion reflecting the relative motion. Some time later there was a suggestion that an Infinite Creator of the universe would have made it infinite in both time and space, and therefore any place could be chosen as its center. These were only speculations, however, and no one undertook the necessary detailed calculations to support such suggestions until the sixteenth century, when a minor church official at the cathedral of Frauenberg (now Frombork, Poland) studied the problem.

This man, Nicolaus Copernicus, studied at universities in Cracow in Poland and Bologna and Padua in Italy, where he acquired a fairly substantial knowledge of mathematics, astronomy, theology, medicine, canon law, and Greek philosophy. His duties at the cathedral did not require much time, so he was able to undertake a prolonged and extensive study of astronomy.

Copernicus ultimately concluded that the Ptolemaic system was too complicated. It now seems clear that Copernicus was influenced by Neoplatonism, a philosophical outlook that had strong roots in classical Greek thought. Consistent with this outlook was a rule called Ockham's razor, which in essence says that simple explanations are preferable to complicated explanations. (In this connection, King Alfonso X of Castille, who in 1252 sponsored a new calculation of astronomical tables using Ptolemy's techniques and additional data, is reported to have complained that God should have consulted him before setting up such a complicated system.) Copernicus was dissatisfied with the inconsistent way in which Ptolemy had applied the various devices discussed above. He regarded the use of the equant as being particularly contradictory to the very idea of perfect circular motion. He therefore reexamined the heliocentric model that had been proposed more than fifteen hundred years earlier by Aristarchus, a Greek astronomer who lived between the times of Aristotle and Ptolemy.

Copernicus, following Aristarchus' suggestions, proposed that the Sun and the fixed stars should be regarded as stationary and the Earth regarded as a planet that circled about the Sun in the same manner and direction as the other five planets. The Moon alone should be regarded as circling about the Earth. The length of time for any planet to complete one round trip around the Sun should be greater the farther the planet from the Sun. Retrograde motion and all the slowing down and speeding up of the planets is simply an optical illusion, which occurs because the direction in which one of the other planets is seen from the Earth depends upon the relative positions of the Earth and the planet in their respective orbits. These relative positions will change over time in the same way the relative positions of racing cars in a long-distance race around an oval track do. A fast planet will increase its lead over a slow planet until it gains a whole "lap" and will pass the slower planet again and again. Therefore as viewed from the Earth the other planets will sometimes seem to be ahead and sometimes behind the Earth. A conjunction is nothing more special than the observation that one planet is about to "lap" another planet.

Figure 2-7 shows schematically how retrograde motion is seen from the Earth. Figure 2-7b shows the relative positions of the Sun, Earth, and Mars at seven different times. The arrows show the directions (compared to the direction of a distant star) in which an observer on Earth would have to look to see Mars at those times. In Figure 2-7c the corresponding directions are superimposed on

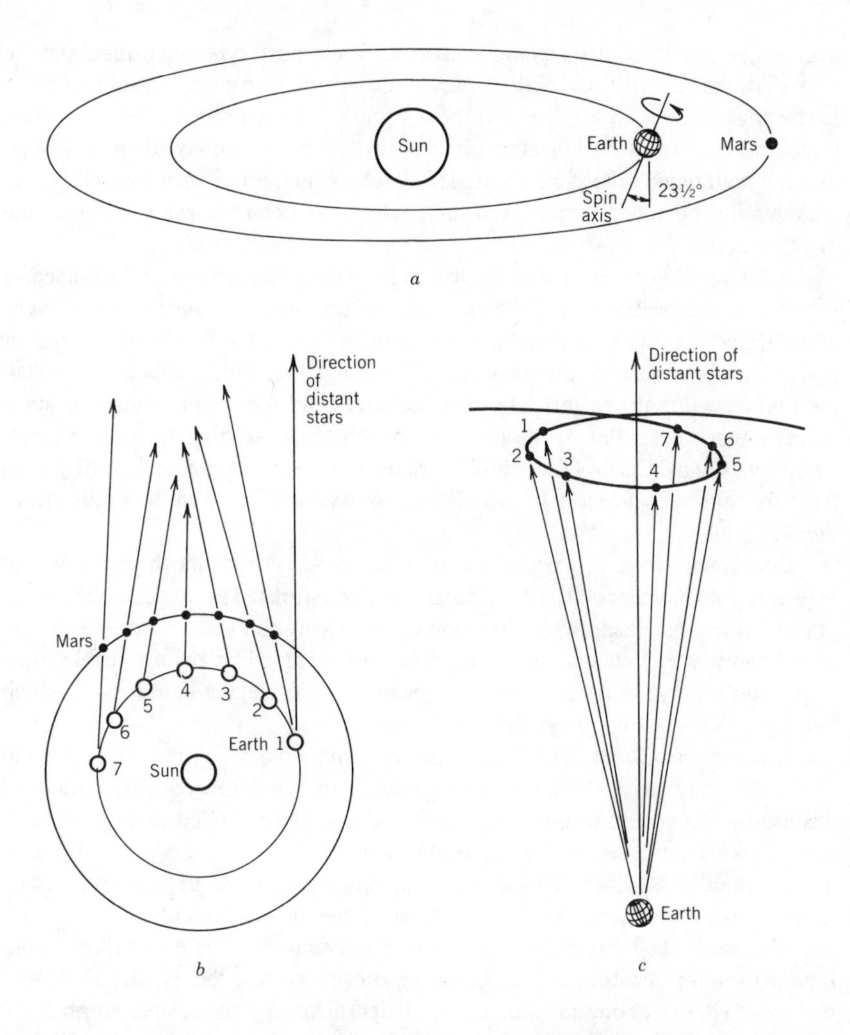

Figure 2-7. Copernicus' heliocentric model. (a) Schematic perspective of Earth and Mars orbiting the Sun, showing the Earth spinning on its axis. (b) Relative positions of Earth and Mars showing different angles of sight from Earth to Mars. (c) Superposition of angles of sight as seen from the Earth.

each other, as they would appear to astronomers or anyone else looking from Earth toward Mars. The lengths of the arrows are adjusted to correspond to the changing distances from Earth to Mars in the seven different positions. The result is a very simple and "natural" explanation of retrograde motion.

To explain the daily rotation (diurnal motion), Copernicus also proposed that the Earth spins on its own axis once each 23 hours and 56 minutes. (The other four minutes to make up the time for an apparent complete rotation of the heavens is needed because the Earth has moved to a different position in its orbit about the Sun.) The spin axis is tilted with respect to the axis of the orbit by an amount equal to the angle (about 23 1/2 degrees) the ecliptic makes with the celestial equator. (Copernicus also asserted that this tilt axis is rotating very slowly about the orbital axis but maintaining the same tilt angle, thereby explaining a very slow change in the overall appearances of the fixed stars, which had been observed since the time of Ptolemy.)

In effect, Copernicus found that by shifting the point of view of the planetary motions from the Earth to the Sun, motions that had heretofore seemed very complicated became very simple: If one could stand on the Sun, it would appear that all the planets are simply traveling in circular paths, with the planets closest to the Sun traveling the fastest. He then deduced from the astronomical measurements the correct order of the planets from the Sun, starting with the closest: Mercury, Venus, Earth, Mars, Jupiter, Saturn. He also calculated their distances from the Sun fairly accurately. The Earth, for example, is 93 million miles from the Sun.

Copernicus then proceeded to calculate exactly how the heavens would appear to astronomers on Earth. In particular, he estimated the appearances of the planets, time and place, for comparison with the astronomical measurements. He found, however, that his calculated positions gave much poorer results than Ptolemaic theory! In other words, his beautiful and elegant conception, which worked so well qualitatively, had failed quantitatively.

Undaunted, Copernicus reintroduced some of the devices Ptolemy had used—the eccentric, deferent, and epicycle (but not the equant, which he despised). Moreover, in introducing these devices, he applied them in a much more consistent manner than was possible using Ptolemy's methods. He also used a much smaller number of spheres of various types: forty-six, compared with more than seventy used in some of the later Ptolemaic-type calculations.

The result of all this effort was astronomical tables that were no better overall than astronomical tables calculated using the geocentric theory and Ptolemy's methods. There was one advantage to the Copernican approach, so far as practical astronomers were concerned: the calculations were somewhat simpler to carry out.

Copernicus completed his work in 1530, but did not publish it in full detail for a number of years, possibly fearing the reaction it might generate; earlier proponents of the idea that the Earth was in motion had been severely ridiculed. Finally, in 1543, on the day that he died, his *On the Revolutions of the Heavenly Spheres* appeared. It was dedicated to Pope Paul III, and its preface (written by someone else) stated that the Copernican scheme should not be taken as representing reality, but rather should be regarded as simply a useful way to calculate the positions of various heavenly objects. The Copernican system and calculations were used in determining the Gregorian calendar, which was first introduced in 1582 and is still in use today.

Initial reaction to the Copernican scheme was largely to regard it in the way suggested by the preface: useful for calculations but not necessarily connected with reality. Copernicus was aware of many of the objections to the heliocentric theory or any theory that involved motion of the Earth since they had been raised as early as the time of the ancient Greeks. In his book he attempted to refute these objections, which can be divided into two categories: (1) scientific objections and (2) philosophic and religious objections. In the minds of most of his contemporaries, the two categories were equally important.

Some of the scientific objections were based on astronomical observations. For example, if the Earth actually moved in a circular orbit about the Sun, the distance from the Earth to a particular portion of the equator of the celestial sphere should vary by as much as 186 million miles (the diameter of the Earth's orbit). The angular separation between two adjacent stars on the sphere should therefore change noticeably, as the angular separation between two trees in a forest changes depending upon how close one is to the trees. This *star parallax* effect had never been observed and Copernicus asserted that the stars were so far away that even a change of 186 million miles was such a small fractional change in the total distance to the stars that the parallax would be too small to be seen by the human eye. His critics rejected this contention because they believed that the celestial sphere was only slightly larger than the sphere of Saturn. (In fact, the nearest stars are so far away that star parallax was not observed until about two centuries later, when sufficiently powerful telescopes became available.)

A related objection involved the apparent size of the stars. Seen with the unaided eye, all starts seem to have the same apparent diameter because they are so far away. If their true diameters are calculated using the distance from the Earth to the stars, asserted by Copernicus, they would turn out to be several hundred million miles in diameter and therefore much larger than the Sun, something also too incredible to believe. As it turns out, the apparent size of the stars as seen with the unaided eye is misleadingly large because of the wave nature of light and the relatively small aperture of the eye; this was not understood until about 300 years later, when the principles of diffraction optics were worked out. (In fact, through a telescope the stars seem smaller than when seen with the naked eye!)

There were also objections based on physics, as the subject was understood at that time. This understanding was based primarily on the writings of Aristotle. For example, it was asserted that if the Earth was truly in motion its occupants should feel a breeze proportional to the speed of the Earth, just as a person riding in an open vehicle feels a breeze. The speed of the Earth as it travels around its orbit is about 67,000 miles per hour. The resulting "breeze" would surely destroy anything on the surface of the Earth (a storm with wind velocities greater than about 75 miles per hour is considered a hurricane).

Similarly, it was claimed that if the Earth is moving, an apple falling from a tree would not simply fall straight down to the ground but would land several miles away because the tree and the ground beneath it would have moved that distance in the time it took the apple to reach the ground. For that matter, how many horses (in modern terms, how big an engine) are needed to pull something as massive as the Earth at such a stupendous speed? There is no evidence of any agent that keeps the Earth moving. Reasoning similarly, a spinning Earth is like a giant merry-go-round, and the resulting centrifugal force should be so large that all the inhabitants would have to hang on for dear life. Because none of these things occur, it was hard to believe that the Earth is actually moving. (The idea that the Earth could exert a gravitational attractive force was not developed until about a century later, as was the distinction between centrifugal and centripetal force.) The proponents of the Copernican system did overcome these scientific objections (see the next chapter), but it took them roughly a century.

The philosophical and theological objections had equal weight. Physics at that time was not accorded the overwhelming respect it has today. Indeed, theology was considered the Queen of the Sciences. Aristotelian philosophy had been integrated into the theological dogma of Western Europe in the sixteenth century. Interestingly enough, a thousand years earlier the Aristotelian outlook had been condemned as pagan and the very idea of a round Earth considered heretical. In Copernicus' time, however, the Aristotelian outlook, which was essentially rational, was dominant, and any flaws in it could be taken as indicative of flaws in the entire accepted order of things. The Earth was considered unique and at the center of the universe, and so too was man.

In Aristotle's thought the Earth and the surrounding sublunar region up to the sphere of the Moon were imperfect. The Moon, Sun, planets, and stars were perfect because they were in the heavens (which was why their motion was circular), perfect objects carrying out the perfect motion that was inherent in their nature. The Earth, being in the sublunar region, was inherently different from the heavenly bodies and could not be a planet like the other planets and therefore could not undergo the same kind of motion as the planets. The planets were also considered to be different because they exerted special influences on individual behavior—individual personalities are still sometimes described as mercurial or jovial or saturnine, for example, traits attributed to the planets Mercury, Jupiter, or Saturn.* A heliocentric theory also weakens the mystic appeal of astrology.

The heliocentric theory was attacked by the German leaders of the Protestant Reformation because it ran counter to a literal reading of the Bible. Certainly the central role of the Sun in the theory could be considered as leading to a return to pagan practices of sun worship. The reaction of the ecclesiastical authorities of the Catholic church was initially rather mild, but ultimately they too condemned heliocentrism as heretical. Some historians have even suggested that if Coper-

*It can be argued that such beliefs are inconsistent with the rationality that should have been imposed by Aristotelian philosophy, but all cultures have inconsistencies. Moreover, medieval Europe had acquired its knowledge of Aristotelian philosophy not from the original Greek texts but from Latin translations of Arabic versions. The translators inevitably added their own ideas to the original text.

nicus had developed his ideas fifty years earlier, well before the onset of the Reformation, they might have received a more sympathetic audience. But coming as they did in the midst of the theological struggles of the sixteenth and seventeenth centuries, they were seen as a challenge to the competent authority of the church, which had long proclaimed a different cosmology (or to the Bible, in the Protestant view), which had to be combated. (Copernicus, and others who supported the heliocentric view, felt that it could be reconciled with the Bible and that the Sun simply represented God's glory and power.)

The challenge perhaps seemed to be even more severe because it led to wild speculations such as those of Giordano Bruno, a monk who proclaimed some fifty years after Copernicus' death that if the Earth were like the planets, then the planets must be like the Earth. They were surely inhabited by people who had undergone historical and religious experiences similar to those that had transpired on Earth, including all the events recorded in the Bible. Bruno asserted that the universe was infinite in extent and the Sun was but a minor star, and that there were other planetary systems. The uniqueness of mankind, of God's concern for mankind and of various sacred practices and institutions was thereby lost. (Bruno was ultimately burned at the stake because he refused to recant and repudiate this and other heresies.)

All things considered, in view of the various objections raised against it and the fact that even after refinement it gave no better results than the Ptolemaic theory, there seemed to be no compelling reason to prefer the Copernican theory. Its chief virtue seemed to be that it was intellectually more appealing to some astronomers and lent itself to easier calculations. It did demonstrate, however, that there was more than one possible scheme for "saving the appearances" and it crystallized the idea that the Ptolemaic theory was no longer satisfactory. The time was at hand for new and better astronomical theories.

New Data and a New Theory

One of the weaknesses of Copernicus' work was his use of large amounts of very ancient, possibly inaccurate, data. He himself made only a limited number of astronomical observations. In 1576, the Danish astronomer Tycho Brahe built an astronomical observatory on an island near Copenhagen under the patronage of the King of Denmark. Brahe was a nobleman of fearsome appearance (he had a silver nose as a result of a duel), but he was an unusually keen and meticulous observer and was able to measure the position of a planet or star quite accurately. He spent the following twenty years making a large number of extremely accurate measurements of the positions of the Sun, Moon, and planets. Where previous measurements had been reliable to an angular accuracy of about 10 minutes of arc, Brahe's data were accurate to about 4 minutes of arc (there are 60 minutes in a degree). To give some idea of what this means, if a yard-long pointer is used to sight a planet, a shift of the end of the pointer by 1/25 inch would give an angular error of 4 minutes of arc. While most previous observers were content to observe the position of planets at certain key points, Brahe tracked them throughout their entire orbit. His observations were not done with telescopes (which were not invented until 1608) but with various "sighting tubes" and pointer instruments he had designed.

Brahe found that neither the Copernican theory nor the Ptolemaic theory agreed with his new data. In fact, he knew even thirteen years before he built his observatory that both theories had serious errors. He observed a conjunction of Jupiter and Saturn, and on checking the astronomical tables found that the Ptolemaic theory had predicted this conjunction would occur a month later, whereas the Copernican theory predicted it would occur several days later. Tycho Brahe was also impressed by the various arguments against the motion of the Earth, and he realized that it was possible to construct yet another theory based on perfect circular motion.

Tycho Brahe
(The N.Y. Public Library Picture Collection)

Brahe therefore proposed that the Earth is at the center of the sphere of the fixed stars and at rest, and that the Sun and Moon travel in orbits around the Earth. The other five planets travel in orbits centered about the Sun. A theory of this type, in which the Earth is still at the center of the universe but in which the other planets revolve about the Sun, is called a Tychonic theory. In Tycho Brahe's version the outer sphere of the stars rotates once in twenty-four hours, giving rise to the diurnal motion. In other Tychonic theories, the stellar sphere is fixed and the Earth rotates daily on its axis while remaining at the center of the sphere.

Tycho Brahe in his observatory
(AIP Niels Bohr Library)

The Tychonic model came to be generally accepted by most astronomers of the time because it preserved something of the intellectual unity of the heliocentric theory and the relative ease of calculations of the Copernican scheme. At the same time Brahe's theory avoided all the difficulties associated with the movement of the Earth. Although this theory was subsequently superseded, it is indeed notable because it represented the recognition of the failure of the Ptolemaic model. Brahe did not work out his theory completely; on his deathbed, he charged his assistant and ultimate successor, Johannes Kepler, with the task of carrying out the detailed calculations necessary.

ARMILLÆ AEQUATORIÆ MAXIMÆ
SESQUIALTERO CONSTANTES CIRCULO

Tycho Brahe's sextant for measuring the positions of planets (AIP Niels Bohr Library)

New Discoveries and Arguments

Copernicus and Brahe were part of a long-standing and continuing tradition in astronomy: the observation and calculation of the positions of heavenly bodies. Starting in the late sixteenth century, however, new categories of information began receiving considerable attention.

In 1572, a new and very bright star was seen in Western Europe. This star flared up and became brighter than any other star and brighter even than the planets; then it faded away from visibility over a period of several months. Careful observations by Brahe and several other astronomers showed that this star was as far away and as fixed in location as any other star. Another similar star was seen in 1604. Such a star, called a nova, is not uncommon. In 1576 a comet was observed moving across the heavens beyond the lunar sphere, and in such a direction that it must have penetrated through several of the planetary spheres.

Such occurrences had been reported in the records of the Greek astronomers, but the quality of the observations had been such that they had been thought to take place in the sublunar region. In 1572, 1576, and 1604 the observations were specially significant; they showed clearly that the phenomena took place in the astronomical or heavenly world. This meant that even the heavens changed. If the heavens change, they cannot be perfect—why else would they change? Yet according to Aristotelian physics, circular motion was reserved to the heavens because of their perfection, whereas the Earth could not be in circular motion because of its imperfection. But the new observations suggested that the heavens were not perfect, so there was no reason why the motion of the Earth should be any different than the motion of a planet.

The observations of novas and comets were just the beginning of a large number of new observations of a qualitatively different kind. In 1608 the telescope was invented. Initially it was intended to be used for military purposes, but in 1609 an Englishman, Thomas Harriot, used one to examine the surface of the Moon. Late the same year, Galileo Galilei, professor of mathematics at the University of Padua in Italy, constructed a greatly improved telescope, which he turned to the heavens. In his hands, the telescope became the source of many significant discoveries. Early in 1610 he published a slim book, the *Sidereus Nuncius* (*Starry Messenger*), describing his observations.

Galileo found that the surface of the Moon was very much like that of the Earth, with mountains and craters whose heights and depths he could estimate, and that the surface of the Sun had spots which developed and disappeared over time. Both of these discoveries indicated that, contrary to Aristotelian philosophy, not all heavenly objects are "perfect." He also found that there were centers of rotation in the universe other than the Earth—the planet Jupiter has moons and the Sun spins on its own axis. He was able to see that the planet Venus went through phases like the Moon's; that is, it could be fully illuminated like a full moon or partially illuminated like a crescent moon. When fully illuminated it was much farther away from the Earth than when it appeared as a crescent. This could be explained only if Venus circled about the Sun.

Galileo was also able to see that there were many more stars in the sky than had been previously realized, stars that were too far away to be seen by the naked eye. Moreover, even though his telescope made objects seem thirty times closer than when seen with the naked eye, the stars remained mere pinpoints of light, indicating that they were indeed at least as far away as Copernicus had claimed. In fact, as already noted, when studied with a telescope, the stars are actually smaller in apparent diameter than when seen with the naked eye because of the greater aperture of the telescope. All this represented an experimental refutation of some of the objections raised against a heliocentric theory.

Earlier Galileo had begun the development of new principles of physics opposed to those propounded by Aristotle with which he was able to counter the scientific opposition to the Copernican system (See Chapter 3). Galileo was a lusty individual, a skilled publicist and polemicist, and participated in the scientific debates of the day. Indeed, he was so devastating in written and oral argument that he made a number of enemies as well as friends.

By 1616, the religious schism arising from the Reformation and the Counter-Reformation had become so great that tolerance of deviations from established doctrines decreased significantly. In that year Galileo was warned not to advocate publicly the Copernican system. In 1623, however, Cardinal Barberini, a notable patron of the arts and sciences, become pope. Thinking that a more favorable climate was at hand, Galileo felt that he could embark upon a more vigorous advocacy of the Copernican system. He ultimately wrote a book, *Dialogues Concerning the Two Chief World Systems*, which was directed to the educated lay public. The style of the book was that of Socratic dialogue, with three characters: an advocate of the Ptolemaic system, an advocate of the Copernican system, and an intelligent moderator. At the end of the book the Copernican advocate suddenly conceded the validity of the Ptolemaic system, although it was quite obvious to the reader that the Copernican system had been shown to be the correct one. In addition, the Ptolemaic advocate was depicted as somewhat of a simpleton.

Somehow Galileo managed to get the manuscript of the book approved by the ecclesiastical censors, and it was printed in 1632. Unfortunately, he had many enemies in high places, and they were quick to point out the true intent and implications of the book. It was also asserted that Galileo intended to ridicule the highest officials of the church. A few months later the book was banned and all remaining copies were ordered destroyed. Galileo was called to trial before the Inquisition and, under threat of torture, compelled to recant his advocacy of Copernican doctrine. Perhaps because of his advanced age, poor health, and submission to the will of the Inquisition, his punishment was relatively light: house arrest and forced retirement. Cared for by his illegitimate daughter, he died nine years later, in 1642, shortly before his seventy-eighth birthday. He no longer actively advocated the Copernican system but devoted his time to writing another book, *Discourses on Two New Sciences*, in which he detailed the results of his researches in mechanics and optics. This book was published in the Netherlands, where the intellectual climate was more hospitable than in Italy.

Kepler's Heliocentric Theory

Both Tycho Brahe and Galileo Galilei used a relatively modern approach to scientific problems. Brahe insisted on the importance of systematic, extensive, and accurate collection of data. Galileo was not content simply to make observations, but recognized the necessity to alter experimental conditions in order to eliminate extraneous factors. Both men were very rational and logical in their approach to their work. Copernicus, on the other hand, and Kepler (of whom more will be said later) even more so, were concerned with the philosophical implications of their work. Kepler was well aware that the ancient Pythagoreans had set great store on finding simple numerical relationships among phenomena. It was the Pythagoreans who recognized that musical chords are much more pleasing if the pitches of their component notes are harmonics—that is, simple multiples of each other. The Pythagoreans also discovered that the pitch of sounds from a taut string are related to the length of the string, the higher harmonics being obtained by changing the length of the string in simple fractions. Turning to the heavens, they asserted that each planetary sphere, as well as the sphere of the stars, emits its own characteristic musical sound. Most humans never notice such sounds, if only because they have heard them continuously from birth. The spatial intervals between the spheres were also supposed to be proportional to musical intervals, according to some of the legends about the Pythagoreans.

Johannes Kepler was born some twenty-eight years after Copernicus' great work was published. He loved to indulge in Neopythagorean speculations on the nature of the heavens. His personality was undoubtedly affected by an unhappy childhood, relatively poor health, and a lack of friends in his youth. He showed mathematical genius as an adolescent, and as a young man succeeded in obtaining positions as a teacher of mathematics. His mathematical ability was soon recognized, and as a result his deficiencies as a teacher and his personality problems were often overlooked. The fact that he was such a poor teacher that he did not have many students allowed him more time for astronomical studies and speculations.

Kepler was convinced of the essential validity of the heliocentric theory and looked upon the Earth as one of the six planets that orbited the Sun. Because of his fascination with numbers, he wondered why there were six planets rather than some other number, why they were spaced as they were, and why they moved at their particular speeds. Seeking a connection to geometry, he recalled that the Greek mathematicians had proved that there were only five "perfect" solids, known as the Pythagorean or Platonic solids, aside from the sphere.

Johannes Kepler
(AIP Niels Bohr Library)

A perfect solid is a multifaced figure, each of whose faces is identical to the other faces. Each of the faces is itself a perfect figure (see Figure 2-8). A cube has six faces, all squares; a tetrahedron has four faces, all equilateral triangles; the eight faces of an octahedron are also equilateral triangles; the twelve faces of a dodecahedron are pentagons; and the twenty faces of the icosahedron are all equilateral triangles. Kepler proposed that the six spheres carrying the planets were part of a nested set of hollow solids in which the spheres alternated with the Platonic solids. If the innermost and outermost solids are spheres, then there could be only six such spheres, corresponding to the number of known planets. The size of the five spaces between the spheres might be determined by carefully choosing the particular order of the Platonic solids, as shown in Figure 2-9.

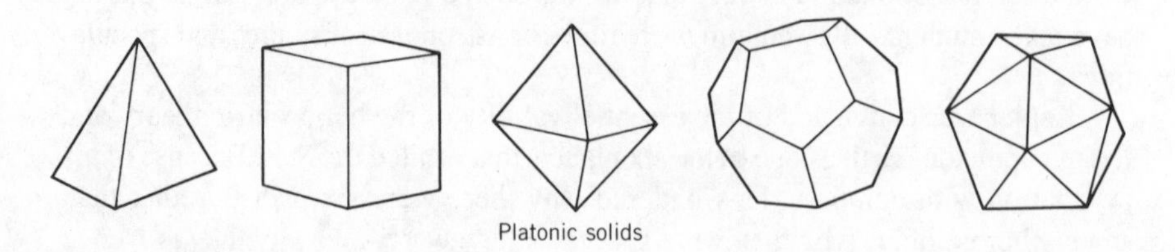

Figure 2-8. The Platonic solids. From left to right: tetrahedron, cube, octahedron, dodecahedron, icosahedron.

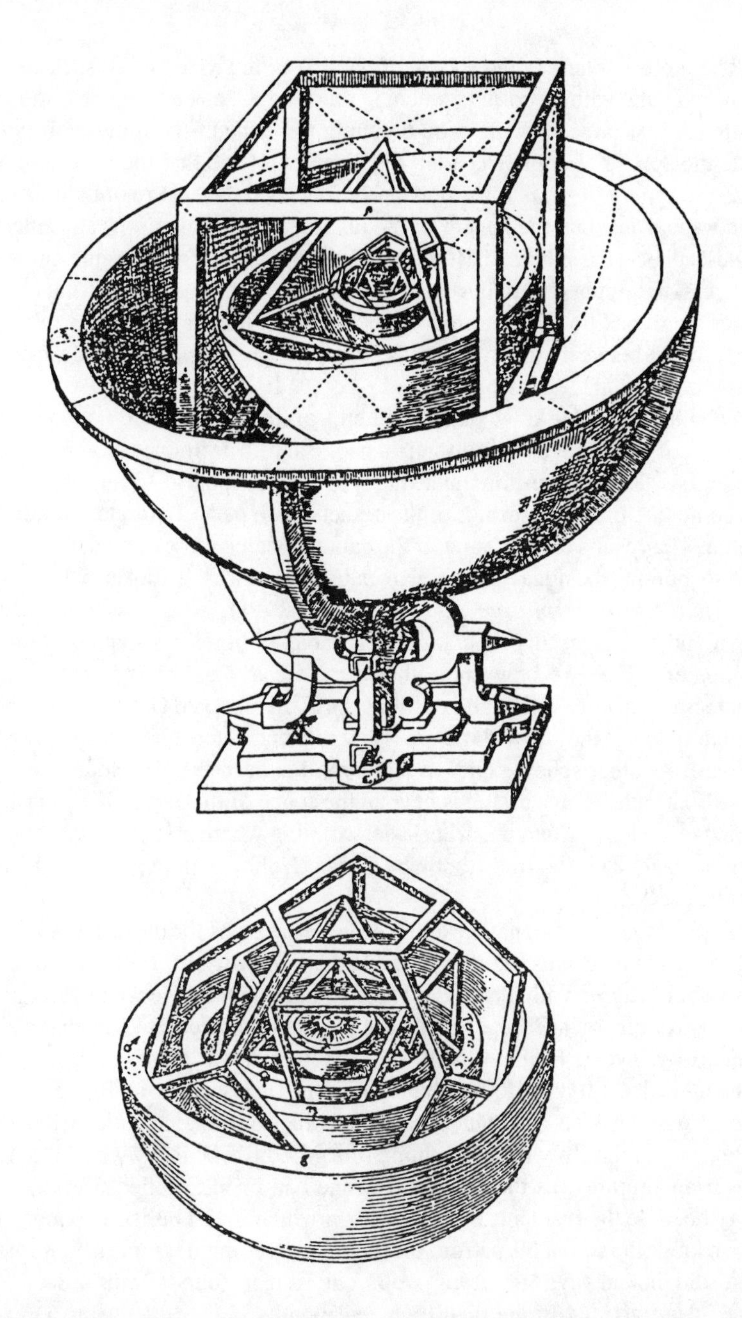

Figure 2-9. Kepler's nesting of spheres and Platonic solids to explain spacings and order of the planets. The lower drawing shows the spheres of Mars, Earth, Venus, and Mercury in more detail. Note the Sun at the center.
(From illustrations in Kepler's *Mysterium Cosmographicum*)

The largest sphere is the sphere of Saturn. Fitted to the inner surface of that sphere is a cube with its corners touching the sphere. Inside the cube is the sphere of Jupiter, just large enough to be touching the cube at the middle of its faces. Inside the sphere of Jupiter is a tetrahedron and inside that the sphere of Mars. Inside the sphere of Mars is the dodecahedron, then the sphere of the Earth, then the icosahedron, then the sphere of Venus, then the octahedron, and finally the smallest sphere, the sphere of Mercury. With this order of nesting, the ratios of the diameters of the spheres are fairly close to the ratios of the average distances of the planets from the Sun. The Earth even retains some uniqueness in this scheme: its sphere rests between the two solids with the greatest number of faces, the dodecahedron and the icosahedron.

Of course, Kepler knew that the spheres of the planets were not homocentric on the Sun; Copernicus had found it necessary to introduce eccentrics in his calculations. Kepler therefore made the spheres into spherical shells thick enough to accomodate the meanderings of the planets from perfect circular motion about the Sun. The results did not quite fit the data, but came close enough for Kepler to want to pursue the idea further. He published this scheme in a book called *Mysterium Cosmographicum* (*The Cosmic Mystery*), which became very well known and estalished his reputation as an able and creative mathematician and astronomer. Since we now know that there are at least nine planets, Kepler's scheme is of no value although questions may still be asked about why the planets have the spacings they do and if there is any rule that determines how many planets there are. Another scheme once thought valid is described by Bode's law or the Titius-Bode rule. A discussion is beyond the scope of this book, but a fascinating account of its history and the criteria determining whether such a rule is useful is given on pages 156–160 of the book by Gerald Holton and Stephen Brush cited in the references.

Kepler was not content to propose esoteric schemes for the order of planets in the heavens and be satisfied if they worked approximately. He believed that any theory should agree with the available data quantitatively. He was aware, as were all the astronomers in Europe, of the great amount of accurate data that had been obtained by Tycho Brahe. Brahe, on the other hand, was aware of Kepler's mathematical ability. In 1600, Kepler became an assistant to Brahe, who had moved to the outskirts of Prague and had become court mathematician to Rudolph II, Emperor of the Holy Roman Empire (the grandiose name by which the Austro-Hungarian Empire was then known). Brahe had been expelled from his observatory because the residents of the island on which it was built complained to the Danish king (the son of his patron) that he ruled over them tyranically. Kepler and Brahe did not always get along well, but Kepler found Brahe's data a very powerful attraction. Brahe died eighteen months later, in 1601 (as a result of overeating), and Kepler immediately took custody of all the data. In 1602 Kepler was appointed Brahe's successor as court mathematician to Rudolph II.

Brahe had urged Kepler to try to refine Brahe's compromise theory to fit the data, but he was not successful. He then returned to the heliocentric theory, but did not share Copernicus' objection to the equant, so he included that in the calculations. He also made some improvements on some of Copernicus' detailed assumptions, making them more consistent with a truly heliocentric theory. He concentrated particularly on the calculations for the orbit of Mars, because this seemed the most difficult one to adjust to the data. The best agreement he could get was, on the average, within 10 minutes of arc. Yet Kepler felt that Brahe's data were so good that a satisfactory theory had to agree with the data within 2 minutes of arc. He felt that the key to the entire problem for all the planets lay in that eight-minute discrepancy. Gradually and painstakingly, over a period of two decades punctuated by a variety of interruptions that included changes of employment, the death of his first wife, and a court battle to defend his mother against charges of witchcraft, he worked out a new theory.

After spending several years studying the data for the planet Mars, Kepler despaired of finding the right combination of eccentric, deferent, epicycle, and equant that would make it possible to describe the orbit in terms of circular motion. He finally decided to abandon the circular shape, saying that if God did not want to make a circular orbit then such an orbit was not mandatory. Kepler then realized that his task was to determine the true shape of the orbit. A few years later, in 1609, he announced as a partial solution to the problem his first two laws of planetary motion. Interestingly enough, Kepler discovered his second law, dealing with the speed of the planet as it travels around the orbit, before he discovered the exact shape of the orbit as given by the first law. Still later, in 1619, he announced the rest of the solution, his third law of planetary motion, which relates the size of the planet's orbit to its speed.

Kepler's laws of planetary motion, as expressed today, are:

I. The orbit of each planet about the Sun is an ellipse with the Sun at one "focus" of the ellipse.

An ellipse can be contrasted with a circle. A circle surrounds a special point called the center, and the distance from the center to any point on the circle is always the same, or constant. An ellipse surrounds two special points, called foci, and the sum of the distances from each focus to any point on the ellipse is always constant. This is illustrated in Figure 2-10a, where the sum of the lengths of the pairs of lines, *A* and *B*, or *C* and *D*, or *E* and *F*, is always the same.

II. The line joining the Sun and the planet (the radius vector) sweeps over equal areas in equal times as the planet travels around the orbit.

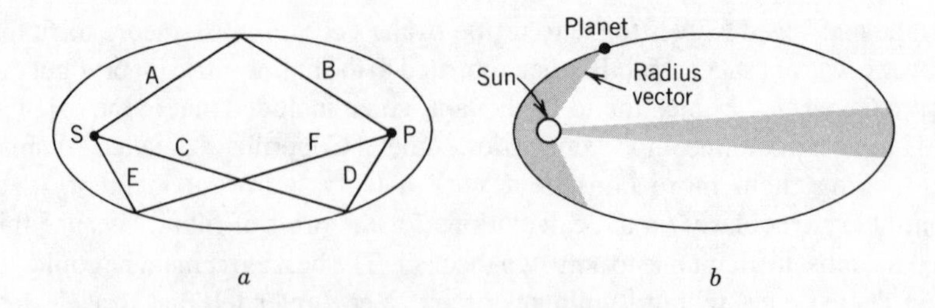

Figure 2-10. Elliptical planetary orbits. (a) Construction of an ellipse: S and P are the two foci. The sum of the distances to any point on the ellipse from the foci ($A + B, C + D, E + F$) is always the same. (b) Kepler's first and second laws. The planet takes the same length of time to traverse the elliptical arcs bounding the two shaded areas if the shaded areas are of equal size.

In Figure 2-10b, the planet takes the same length of time to travel along the elliptical arcs bounding the two shaded areas because the areas are the same size. Thus when the planet is close to the Sun, its speed increases, and when it is far from the Sun, its speed decreases.

> *III. The square of the period of revolution (time for one complete orbit) of a planet about the Sun is proportional to the cube of the average distance of the planet from the Sun.*

Mathematically, this means that the ratio of T^2 to D^3, where T stands for the period and D the average distance, is the same for all planets. The third law is illustrated in Table 2-1, which gives the period T (in years) of the planets; their average distance, D (in astronomical units, with one astronomical unit equal to 93 million miles); the period squared, T^2; and the average distance cubed, D^3. With these units, the ratio is equal to one—that is, $T^2 = D^3$.

With these three laws Kepler completed the program set forth by Copernicus. Copernicus made the Sun the center of the planetary system; Kepler discarded the idea that circular motion was required for heavenly bodies. By using elliptical orbits, the true beauty and simplicity of the heliocentric theory was preserved, in excellent agreement with the data and with no need for such devices as the eccentric, epicycle, or equant. The fact that the focus of an ellipse is not in the center supplies eccentricity. Kepler's second law performs the same function as an equant in that it accounts for the varying speed of the planet as it travels around the orbit.

Kepler went further and pointed to the Sun as being the prime mover or causative agent, which he called the *anima motrix*. He put the cause of the motion at the center of the action rather than at the periphery, as Aristotle had done. Kepler had a primitive idea about gravitational forces exerted by the Sun but failed to connect gravity to planetary orbits. Rather he asserted that the Sun exerted its

Table 2-1 Data Illustrating Kepler's Third Law of Planetary Motion for the Solar System

Planet	T	D	T^2	D^3
Mercury	0.24	0.39	0.058	0.059
Venus	0.62	0.72	0.38	0.37
Earth	1.00	1.00	1.00	1.00
Mars	1.88	1.53	3.53	3.58
Jupiter	11.9	5.21	142	141
Saturn	29.5	9.55	870	871

influence through a combination of rays emitted by the Sun and the natural magnetism of the planet. He was influenced in this belief by some clever demonstrations of magnetic effects by William Gilbert, an English physician and physicist.

The complete translated title of the book he published in 1609 on the first two laws is *A New Astronomy Based on Causation, or a Physics of the Sky Derived from Investigations of the Motions of the Star Mars, Founded on Observations of the Noble Tycho Brahe* (usually referred to as the *New Astronomy*). Kepler had actually completed an outline of the book in 1605, but it took four years to get it printed because of a dispute with Tycho Brahe's heirs over the ownership of the data.

Kepler's third law, called the Harmonic Law, was described in a book called *Harmony of the World*, published while he held a new and lesser position as provincial mathematician in the town of Linz, Austria (his former patron, Rudolph II, had been forced to abdicate his throne, and Kepler lost his job). In this book Kepler returned to the Pythagorean mysticism of his early days, searching for harmonic relationships among the distances of the planets from the Sun. Instead of finding a relationship between the "musical notes" of the heavenly spheres and their size, he found the relationship between the speed of the planets (that is, their periods), which he somehow associated with musical notes and harmony, and the size of their orbits. He supported the connections with harmony by some surprisingly accurate calculations.

Kepler was the right person at the right time to discover the laws of planetary motion. First and perhaps most significantly, Kepler was an intense Pythagorean mystic. He was also an extremely capable mathematician who believed that the universe was full of mathematical harmonies. Because he was so interested in figures and shapes, it was relatively easy for him to consider elliptical orbits —once he finally realized that circular constructions must be wrong. Finally, Kepler was Brahe's assistant, so that he had access to Tycho Brahe's measurements and knew how accurate they were. No one else would have had the ability or the interest to perform all the mathematical calculations required to discover the laws of planetary motion, and no one else would have regarded an eight-minute angular discrepancy as significant.

Kepler sent copies of his work to many of the best-known astronomers of his day and corresponded copiously with them. In general, although he was greatly respected, the significance and implications of his results remained unrecognized for a long time except by a few English astronomers and philosophers. Most astronomers had become so accustomed to the relatively poor agreement between theory and data, and so imbued with the idea of circular motion, that they were not much impressed with Kepler's results.

Galileo, for example, totally ignored Kepler's work even though they had corresponded and Galileo was quick to seek Kepler's approval of his own telescopic observations. It is not clear whether Galileo was put off by Kepler's Pythagorean mysticism or so convinced of circular motion as the natural motion for the Earth and other planets that he simply could not accept any other explanation. The fact is, however, that Galileo did not embark on his telescope work until after Kepler published the *New Astronomy*, nor did he publish his *Dialogues* until 1632, two years after Kepler's death.

Only after Kepler calculated a new set of astronomical tables, called the *Rudolphine Tables* and published in 1627, was it possible to use his ideas and the data of Tycho Brahe to calculate calendars that were much more accurate than any previously calculated. For example, discrepancies in tables based on the Copernican theory had been found by Tycho Brahe within about ten years of their publication; the *Rudolphine Tables* were used for about a century. During that time Kepler's work gradually became more and more accepted but not until some seventy years later, when Isaac Newton published his laws of motion and his law of universal gravitation in order to explain how Kepler's laws had come about, did they become completely accepted.

The Course of Scientific Revolutions

The replacement of the geocentric theory of the universe by the heliocentric theory is often referred to as the Copernican revolution. As a "revolution" the Copernican revolution is the archetype of scientific revolution, against which other scientific revolutions are compared. In fact, no other scientific revolution can really match it in terms of its effect on modes of thought, at least in Western civilization. The only one that comes close to it in that respect is the Darwinian revolution, which is associated with the development of biological evolutionary theory.

The Copernican revolution is often pointed to as illustrating the conflict between science and religion, yet its leaders—Copernicus, Galileo, and Kepler—were devout men. On the other side, within the religious establishment, there were many who did not follow the narrowly literal interpretation of religious dogma. Initially at least, the biggest fear of the revolutionaries was not censure by religious authorities but ridicule by their scientific peers. How could they challenge the established body of scientific knowledge and belief?

The revolution, moreover, did not take place overnight. The time span from the serious studies of Copernicus to Newton's publication of his gravitational studies was about 150 years. The seeds of the revolution went back some two thousand years before Copernicus to Philolaus, a Pythagorean. As the older theories began to accumulate more and more anomalous and unexplained facts, the burden of necessary modifications simply became overwhelming, and the older theories broke down. Nevertheless, those older theories had been phenomenally successful. In fact, only a few heavenly bodies out of the whole host of the heavens (the planets) caused major difficulties for established theory. The revolution came about because of quantitative improvements in the data and the recognition of subtle rather than gross discrepancies, not only in the data but also in the way in which the data were assimilated. The development of new technologies and instruments that yielded entire new classes and qualities of data also played a large role. At the same time, the new concepts themselves initially could not account for all the data, and their advocates were required to be almost unreasonably persistent in the face of severe criticism. A new theory is expected, quite properly, to overcome a large number of objections, and even then some of the advocates of the old theory are never really convinced.

Not all revolutionary ideas in science are successful, or even deserve to be. If they fail to explain the existing facts or future facts to be developed, or if they are incapable of modification, development, and evolution in the face of more facts, they are of little value. The heliocentric theory was successful because it was adaptable. Kepler, Copernicus, and others were able to incorporate into the scheme such things as eccentrics and epicycles, and finally a different-shape orbital figure, the ellipse, without destroying the essential idea, so that it could be stretched to accommodate a wide variety of experimental facts.

Isaac Newton
(Photo courtesy Yerkes Observatory, University of Chicago)

3 Newtonian Mechanics and Causality

The universe is a mechanism run by rules

It was the development of the branch of physics known as mechanics that finally explained the laws of planetary motion discussed in Chapter 2. The science of mechanics, concerned with the description and causes of the motion of material objects, explains the motions not only of celestial bodies but also of earthly objects, including falling bodies and projectiles. Mechanics is involved to some degree in all physical studies and is sometimes called the backbone of physics. After Isaac Newton, it seemed that the laws of mechanics could explain nearly all of the basic physical phenomena of the universe. This success was widely acclaimed and contributed significantly to the development of the Age of Reason, a time when many scholars believed that all of the sciences, including the social sciences and economics, could be explained in terms of a few basic underlying principles.

Aristotelian Physics

Among the earliest contributors to the science of mechanics were the ancient Greek philosophers, particularly Aristotle. The Greeks inherited various strands of knowledge and ideas concerning natural phenomena from previous civilizations, especially from the Egyptian and Mesopotamian cultures. Although a large amount of knowledge had been accumulated (considerably more than most people realize), very little had been done to understand this information. Explanations were generally provided by various religions, but not by a series of rational arguments starting from a few basic assumptions. Thus, for example, the early Egyptians said the Sun traveled across the sky because it rode in the Sun god's chariot. Such an explanation clearly does not need a logical argument to support its conclusion.

It was the early Greeks who began to try to replace religious explanations with logical arguments based on simple underlying theories. Their theories were not always correct, but their most important contribution was the development of an approach for explaining physical phenomena. The essential objective of this approach was clearly stated by Plato in his Allegory of the Cave, discussed in Chapter 2. The goal of science (or natural philosophy as it was previously called) is to "explain" the appearances—that is, to explain what is observed in terms of the basic principles.

Perhaps the first person to make a serious attempt to develop a unified physical theory was Aristotle (384–322 B.C.), a student of Plato and teacher to Alexander the Great. So far as is known, Aristotle was the first philosopher to set down a comprehensive system based on a set of simple assumptions from which he could rationally explain all of the known physical phenomena. Chapter 2 gives an introduction to his system in connection with the movements of the heavenly bodies. Aristotle wished to explain the basic nature and properties of all known material bodies; these properties included their weights, hardnesses, and motions. Indeed, his descriptions of the physical universe were matter-dominated. He began from only a few basic assumptions, then proceeded to try to explain more complicated observations in terms of these simple assumptions.

Aristotle believed that all matter was composed of varying amounts of four prime substances: earth, water, air, and fire. He believed that objects in the heavens were composed of a heavenly substance called ether, but Earthly objects did not contain this heavenly substance. Thus most of the heaviest objects were regarded as earth and water, whereas lighter objects included significant amounts of fire or air. The different characteristics of all known substances were explained in terms of different combinations of the four primary substances. Furthermore, and more importantly for his development of the subject of mechanics, Aristotle believed that motions of objects could be explained as due to the basic natures of the prime substances.

Aristotle enumerated four basic kinds of motion that could be observed in the physical universe: (1) alteration, (2) natural local motion, (3) horizontal or violent motion, and (4) celestial motion. Alteration (or change) is not regarded as motion at all in terms of modern definitions. For Aristotle, the basic characteristic of motion is the fact that the physical system in question changes. Thus he regarded rusting of iron, leaves turning color, or colors fading a kind of motion. Such changes are now considered part of chemistry. We now consider motion to mean

Bust of Aristotle
(Art Reference Bureau)

the physical displacements of an object, and rusting or fading involves such displacements only at the submicroscopic level. The understanding of these kinds of "motion" is beyond our scope and purpose here, although some aspects of molecular and atomic motions will be discussed later.

The second of Aristotle's categories of motion, natural local motion, is at the center of his basic ideas concerning the nature of motion. For him, natural motion was either "up" or "down" motion. (The Greeks believed that the Earth is a sphere and that down meant toward the center of the Earth and up meant directly away from the center of the Earth.) Aristotle knew that most objects, if simply released with no constraints on their motion, will drop downward, but that some things (such as fire, smoke, and hot gases) will rise upward instead. Aristotle considered these downward and upward motions natural and a result of the dominant natures of the objects, because the objects did not need to be pushed or pulled. Thus earth (rocks, sand, and the like) was said to want to move "naturally" toward the center of the Earth, the natural resting place of all Earthly material. Water was likewise believed "naturally" to want to move toward the center of the Earth. Only at the center of the Earth would earthly material no longer move naturally.

Objects dominated by either of the prime substances fire or air, however, would naturally rise, Aristotle believed. This was explained also as being due to the nature of air and fire: the natural resting place for these substances was in the heavens. Thus air and fire, when liberated, would "naturally" rise. Aristotle viewed all "natural" motion as being due to objects striving to be more "perfect." He believed that earth and water became more perfect as they moved downward toward their natural resting place and that fire and air became more perfect as they moved upward toward their natural resting place.

Aristotle did not stop with his explanation of why objects fall or rise when released, but went further and tried to understand how different sorts of objects fall relative to each other. He studied how heavy and light objects fall and how they fall in different media, such as air and water. He concluded that in denser media objects fall more slowly and that heavy objects fall relatively faster than light objects, especially in dense media. He realized that in less dense media (less resistive, we would say) such as air, heavier and lighter objects fall more nearly at the same speed. He even correctly predicted that all objects would fall with the same speed in a vacuum. However, he also predicted that this speed would be infinitely large. Infinite speeds would mean that the object could be in two places at once (because it would take no time at all to move from one place to another), he argued—which is absurd! Therefore he concluded that a complete vacuum must be impossible ("Nature abhors a vacuum"). As we will see, this last conclusion was later the center of a long controversy and caused some difficulties in the progress of scientific thought.

Although not all of Aristotle's conclusions regarding the nature of falling objects were correct, it is important to understand that his methods constituted at least as much of a contribution as the results. Aristotle relied heavily on careful observation. He then made theories or hypotheses to explain what he saw. He refined his observations and theories until he felt he understood what happened, and why.

Aristotle's third kind of motion was horizontal motion, which he divided into two basic types: (1) that of objects that are continually pushed or pulled and (2) thrown or struck objects, or projectiles. He considered horizontal motion unnatural; that is, it did not arise because of the nature of the object and would not occur spontaneously when the object was released. The first type, which includes such examples as a wagon being pulled, a block being pushed, or a person walking, did not present any difficulty. The cause of the motion appears to reside in the person or animal doing the pulling, pushing, or walking.

The second type of horizontal motion, projectile motion, was harder for Aristotle to understand and represents one of the few areas of physical phenomena in which he seemed somewhat unsure of his explanations. The difficulty for him was not what caused the projectile to start its motion (that was as obvious as the source of pushed or pulled motion) but what made the projectile keep going after it had been thrown or struck. He asked the simple question "Why does a projectile keep moving?" Although this seems to be an obvious and important question, as we will see, there is a better and more fruitful question to ask about projectile motion.

Aristotle finally arrived at an explanation for projectile motion, although one senses that he was not completely convinced by his own solution. He suggested that the air pushed aside by the front of a projectile comes around and rushes in to fill the temporary void (empty space) created as the projectile moves forward and that this inrushing air pushes the projectile along. The act of throwing or striking the object starts this process, which continues by itself once begun. This entire process came to be known as *antiperistasis* and was the first part of Aristotle's physics to be seriously challenged. Aristotle himself wondered if this were really the correct analysis and therefore also suggested that perhaps the projectile motion continued because a column of air was set into motion in front of and alongside the object during the throwing process. This moving column of air, he suggested, might then "drag" the projectile along with it. This bit of equivocation was unusual for Aristotle and emphasizes his own doubts about his understanding of projectile motion.

Aristotle's final kind of motion was celestial or planetary motion. He believed that objects in the heavens were completely different from Earthly objects. For him heavenly objects were massless perfect objects made up of the celestial ether. He accepted the Pythagorean idea that the only truly perfect figure is a circle (or a sphere in three dimensions) and thus argued that these perfect objects were all spheres moving in perfect circular orbits (see Chapter 2). Remember that Aristotle considered heavenly objects and motions to obey different laws than Earthly, imperfect objects. He regarded the Earth as characterized by imperfection in contrast to the heavens and that the Earth was the center of imperfection because things were not in their natural resting places. The Earthly substances earth and water had to move downward to be in their natural resting place, and air and fire had to move upward to be in theirs. The Earth was imperfect because all these substances were mixed together and not in their natural locations.

Because Aristotelian physics has now been largely discarded, we tend to underestimate the importance of Aristotle's contributions to science. His work represents the first "successful" description of the physical universe in terms of logical arguments based on a few simple and plausible assumptions. That his system was reasonable, and nicely self-consistent, is testified to by the fact that for nearly two thousand years it was generally regarded as the correct description of the universe by the Western civilized world. His physics was accepted by the Catholic church as dogma (Aristotle's physics was still being taught in certain Catholic schools and in some Islamic universities in the twentieth century).

But much more important than the actual system he devised was the method he introduced for describing the physical universe. We still accept (or hope) that the correct description can start from a few simple assumptions and then proceed through logical arguments to describe even rather complicated situations. Important also were the questions he considered: What are the objects made of? Why do objects fall? Why do the Sun, Moon, and stars move? Why do projectiles keep moving? Aristotle, more than any other individual, first phrased the basic questions of physics.

With the decline of Greek political power and the rise of Rome, scientific progress in the ancient world slowed to a crawl. Although the Roman Empire gave much to Western civilization in certain areas (including law, engineering, and architecture), it reverted largely to the earlier practice of explaining physical phenomena in religious terms. What scientific progress there was (and there was little compared with the large contributions of the Greek philosophers) came from the eastern Mediterranean. Alexandria, Egypt, was the site of a famous library and a center for scholars for several hundred years. It was in Alexandria that Claudius Ptolemy (some five hundred years after Aristotle) compiled his observations of the wandering stars and devised his complex geocentric system, discussed in Chapter 2.

Another three hundred fifty years after Ptolemy, in the year 500, also in Alexandria, John Philoponus (John the Grammarian) expressed some of the first recorded serious criticism of Aristotle's physics. John Philoponus criticized Aristotle's explanation of projectile motion (the notion of antiperistasis), arguing that a projectile acquires some kind of motive force as it is set into motion. This is much closer to our modern understanding, as we shall see, that the projectile is given momentum that maintains it in motion. Philoponus also questioned Aristotle's conclusion that objects of different weights would fall with the same speed only in a vacuum (which Aristotle rejected as being impossible). Philoponus demonstrated cases in which objects of very different weights fell with essentially the same speeds in air. It is remarkable that these first serious criticisms came more than eight hundred years after Aristotle's death.

With the fall of the Roman Empire and the beginning of the Dark Ages in Europe, and in particular with the destruction of the library and scholarly center in Alexandria, essentially all scientific progress ended. Some work, especially in astronomy, was carried on by Arabian scholars, who also preserved much of Greek philosophy. Finally, another seven hundred years later, the Renaissance began and European scholars rediscovered the early Greek contributions to philosophy and art. One of the main thrusts of the Renaissance was a recommitment to Greek modes of thought. Much of the early Greek work had to be obtained from Arabic texts, principally in Spain and Sicily, where Christianity and Islam had overlapped. In the thirteenth century, St. Thomas Aquinas (1225–1274) adroitly reconciled the philosophy of Aristotle to Catholic doctrine. By the beginning of the fourteenth century, Aristotle's philosophical system had become church dogma and thereby a legitimate subject for study. In the process, fundamental questions regarding the nature of motion were raised again, particularly by scholars at the University of Paris and at Merton College in Oxford, England. Jean Buridan (1295–1358) and William of Ockham (1285–1349) were noted students of motion and became serious critics of some of Aristotle's ideas. Buridan resumed Philoponus' attack on Aristotle's explanations of projectile motion. Buridan provided examples of objects for which, he emphasized, the explanation of antiperistasis clearly would not work; for example, a spear sharpened at both ends (how could the air push on the back end?), and a millwheel (with no back end at all). Buridan also rejected the alternate explanation that a column of moving air was created in front of and alongside a projectile that kept the object moving. He noted that if this were true, one could start an object in motion by creating the moving air column first—and all attempts to do this fail. Buridan, like Philoponus, concluded that a moving object must be given something which keeps it moving. He called this something *impulse* and believed that as an object kept moving it continually used up its impulse. When the impulse originally delivered to the object was exhausted, it would stop moving. He likened this process to the heat gained by an iron poker placed in a fire. When the poker is removed from the fire, it clearly retains something that keeps it hot. Slowly whatever is acquired from the fire is exhausted, and the poker cools. As we will see, Buridan's impulse is actually quite close to the correct description of the underlying "reality" eventually provided by Galileo and Newton.

There was other criticism of Aristotle's physics. In 1277, a religious council meeting in Paris condemned a number of Aristotelian theses, including the view that a vacuum is impossible. The council concluded that God could create a vacuum if He so desired. Besides the specific criticisms by Buridan, scholars of this era produced, for the first time, precise descriptions and definitions of different kinds of motion and introduced graphical representations to aid their studies. These contributions, chiefly by the Scholastics in Paris and the Mertonians in England, provided the better abstractions and idealizations that made possible more precise formulations of the problems of motion. This ability to idealize and better define the problem was essential for further progress. For example, it was at this time that the specific kinds of motion were first understood in an explicit way. *Uniform motion* was defined as motion in a straight line at constant speed. Difform or nonuniform motion, now called *accelerated motion*, was understood to correspond to motion with either changing speed or direction. *Uniformly accelerated motion* (they called it uniformly difform motion) is motion that changes speed at a constant rate; for example, a car that speeds up exactly 5 mph every ten seconds would be moving with uniformly accelerated motion. These careful definitions made possible more precise descriptions of motion, including the motion of falling objects and of projectiles. Especially helpful was the development of graphs to depict motion. For the first time, scholars could make representations of motion that could be analyzed and classified. The two basic kinds of motion are shown in such a graphical representation in Figure 3-1. The graph is a plot of speed versus time. The horizontal straight line represents a constant speed; this is uniform motion. The straight line with slope represents accelerated motion; the speed is continually increasing with time. Actually, because it is a *straight* line, the speed is changing at a constant rate and the sloped line represents uniformly accelerated motion.

These developments were extremely important because they made it possible, for the first time, to analyze motion precisely and in detail, whether of falling bodies or projectiles, in terms of specific kinds of motion. The definitions of uniform and uniformly accelerated motions were critical to properly describe motion, so that the underlying nature of the motions could be discovered. The graphical representations were needed in order to properly recognize these different types of motion in experimental measurements. The stage was thus finally set for someone to determine precisely the kinds of motion involved with falling bodies and projectiles.

Galilean Mechanics

Galileo Galilei (1564–1642) first determined exactly with what kind of motion objects fall. Galileo was born and raised in northern Italy just after the height of the Italian Renaissance. This was a time of religious and political strife and of intellectual and literary activity in all of Europe: Reformation and Counter-Reformation, the Spanish Armada and the reign of Queen Elizabeth I of England, the essays of Montaigne in France and the plays of Shakespeare in England. Galileo demonstrated unusual abilities in several areas as a young man (he probably could have been a notable painter) and studied medicine at the University of Pisa, one of the best schools in Europe. But he early became more interested in mathematics and natural philosophy (in modern terms, science) and displayed so

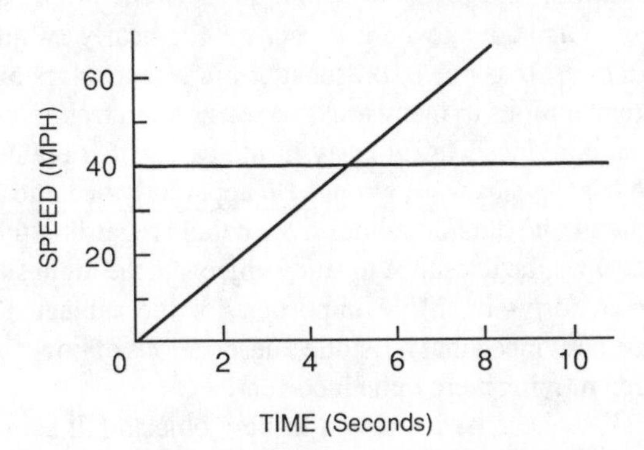

Figure 3-1. Uniform and uniformly accelerated motion indicated on a speed-versus-time graph.

Galileo
(Courtesy of Editorial Photocolor Archives)

much ability that he was appointed professor of mathematics at Pisa while still in his mid twenties. Following some difficulties with the other faculty at Pisa (he was aggressive, self-assured, and arrogant), Galileo at twenty-eight moved to the chair of mathematics at Padua, where he stayed for nearly twenty years. He quickly established himself as one of the most prominent scholars of his time and made important contributions to many areas of science and technology.

Galileo became convinced fairly early in his career that many of the basic elements of Aristotle's physics were wrong. He apparently became a Copernican rather quickly, although he did not publicize his beliefs regarding the structure of the heavens until he used the telescope to study objects in the night sky about 1610, when he was almost forty-six. More important for the subject of mechanics, Galileo also became convinced that Aristotle's descriptions of how objects fall and why projectiles keep moving were quite incorrect.

Apparently Galileo became interested in how objects fall as a young man, while still a student at Pisa, although it is probably not true that he performed experiments with falling objects from the Leaning Tower. He realized that it was necessary to "slow down" the falling process so that one could accurately measure the motion of a falling object. Galileo also realized that any technique devised to solve this problem must not change the basic nature of the "falling" process.

Galileo rejected the Aristotelian approach of observing the slow fall of objects in highly resistive media such as water and other liquids. From such studies Aristotle concluded that heavy objects always fall faster than lighter ones and that objects fall with uniform (constant) speeds. Aristotle noted that heavy and light objects fall at more nearly the same speeds in less resistive media, but felt they would have the same speeds only in a vacuum, which he rejected as being impossible to obtain. He therefore felt that the medium was an essential part of the falling process.

Galileo, on the other hand, felt very strongly that any resistive effects would mask the basic nature of falling motion. He believed that one should first determine how objects fall with no resistive effects, then consider separately the resistive-medium effects, and combine the two togehter to get the net result. The important point here is that Galileo felt the basic nature of the falling process did not include these resistive-medium contributions. Thus he wanted to find some other way to slow falling without changing the basic nature of the process.

Galileo focused on two cases he felt satisfied his criteria. The first was the pendulum, which he first studied as a student. The second was that of a ball rolling down an inclined plane. In both cases he believed that the underlying cause of the motion was the same as that responsible for falling (gravity). Galileo argued that the motions involved should have the same basic characteristics but should be somewhat diminished for pendulums and rolling balls on inclined planes. He obtained the same conclusions from his studies of both cases.

Let us consider his second case, the inclined plane, in some detail. By choosing an inclined plane without too much slope, one can slow down the process in order to observe it. Galileo did more than just observe the motion of the rolling ball, however; he carefully studied how far the ball would travel in different intervals of time. He was able to draw a graph of distance versus time for the rolling ball because he knew how to identify the kind of motion involved from such a graph. (Galileo claimed to be the first to define uniform and uniformly accelerated motions, although he was not.)

It is important to understand that at that time such measurements were not easy. Galileo's success came from his skills and genius as an experimentalist as well as the incisiveness of his thought. For example, there were no accurate stopwatches, so Galileo had to devise ways to measure equal intervals of time accurately. He reportedly started measuring time intervals simply by counting musical beats maintained by a good musician. His final technique was to use a water clock, which measured time by how much water (determined by weight) would accumulate when allowed to drain at a fixed rate. His inclined planes were several feet long and uniformly straight and smooth throughout their length.

The results of Galileo's inclined-plane experiments are represented in a simplified form in Table 3-1 and plotted in Figure 3-2. The distance traveled at the end of the first unit of time is taken to define one unit of distance. The distance traveled increases with the square of the time elapsed. Thus, at the end of two units of time, the ball has rolled four units of distance; at the end of three units of time, it has rolled nine units of distance, and so on. Galileo was able to recognize this progression (and the corresponding graph) as indicative of uniformly accelerated motion. For example, the data of Table 3-1 can be converted to a speed-versus-time graph similar to that discussed in connection with Figure 3-1. If this is done, one obtains Figure 3-3, which shows a straight line with slope. As discussed earlier, this is the result one obtains for uniformly accelerated motion. The speed increases at a uniform rate. Thus Galileo was able to determine experimentally how objects fall: with uniformly accelerated motion. This is the kind of motion a falling object has when the effects of the resistive medium can be eliminated and is the kind of motion that characterizes the basic falling process.

Galileo certainly was aware that falling objects, in air or in liquids, do not keep falling faster and faster; that is, he knew that the resistance of the medium will eventually cause the acceleration to cease and that the falling velocity will reach a constant value. This final maximum velocity is now generally referred to as the terminal velocity for a falling object. Galileo also knew that heavier (denser) objects normally fall faster through the air than lighter (less dense) objects. (This is exactly opposite to the conclusion he is often purported to have reached from experiments at the Leaning Tower of Pisa.) The point is that Galileo realized that this result was from the "interfering" resistance of the air. It was precisely because of the resistive effects of the air (or a liquid) that Galileo decided he needed to study falling with pendulums and inclined planes, thereby keeping the speed low

Table 3-1 Total Distance Traveled after Different times for the Inclined Plane Experiment

Time Elapsed	Distance Traveled
0	0
1	1
2	4
3	9
4	16
-	-
-	-
-	-

enough to minimize the effects of the medium. In order to understand the net result of the falling process, let us consider Figure 3-4.

The figure represents the velocity (speed) of both a heavy and a light object as a function of time. The diagonal line represents the motion of an object falling with a constant (uniform) acceleration, which is the motion an object would have when falling in a vacuum. In anything but a vacuum the resistance of the medium increases as the speed increases until the resistive force equals the downward force of gravity, and the acceleration stops. The resulting terminal velocity for an object will depend not only on its weight but also on its size and shape as well. The terminal velocity for a falling person (a skydiver, for example) may be as much as 130 mph, depending upon body position, clothing, and so on. The horizontal lines in Figure 3-4 indicate the kind of motion predicted by Aristotle for a light and a heavy object. These lines represent constant (although different) speeds, with no indication of a time during which the object speeds up from zero speed to its final terminal velocity.

Although it is important to understand the actual motion of a falling object, as shown in Figure 3-4, from a scientific viewpoint it is more important to understand the true nature (the "true reality" in Plato's Allegory of the Cave) of falling, disregarding the resistance of the medium. Galileo asserted that this true nature is that objects should fall with uniformly accelerated motion as shown in Figure 3-3. This was a very important step toward understanding the falling process. Once the "how" of falling motion is known, it is then necessary to know "why" things fall. The understanding of why was provided by Isaac Newton, who was born about a year after Galileo died. Before discussing Newton's contributions, however, let us consider some other important contributions to the science of mechanics by Galileo and also the general relationship of mathematics to science.

As part of Galileo's studies of objects rolling down inclined planes, he was able to arrive at two very important generalizations regarding motion in general. Both of these new insights applied directly to Aristotle's old problem of how to explain projectile motion.

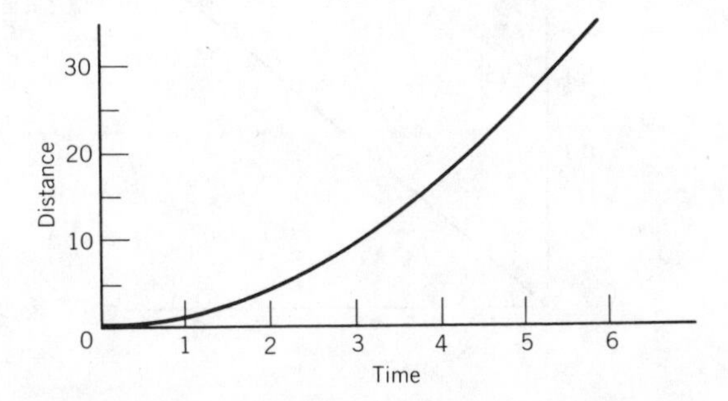

Figure 3-2. Graphical representation of distance versus time for Galileo's inclined-plane experiments. The data from Table 3-1 fit this graph.

Galileo noted that a ball rolling down an inclined plane would continually experience uniform acceleration (Table 3-1), even for a very small angle of inclination. He then wondered what would happen to a ball rolling *up* an inclined plane. He discovered that a ball started along a flat surface with uniform motion (constant speed) would experience uniform deceleration (that is, *negative acceleration*) if it then rolled up an inclined plane. Because these two conclusions appeared always to be true, even for very small angles of inclination, he concluded that as the angle of the plane was changed from slightly downward to slightly upward, the acceleration must change from slightly positive to slightly negative, and an angle must exist for which there is no acceleration. This angle, he concluded, must correspond to a horizontal plane, with no slope whatsoever.

Galileo's analysis was extremely important because, as he realized, if a ball rolling on a flat plane experienced no acceleration or deceleration, it would naturally roll along forever. That is, Galileo realized that a ball rolling along a perfectly level surface would naturally keep rolling with uniform speed unless something (such as friction or upward inclined slopes) acted on the ball to cause it to slow down. Thus Galileo recognized that Aristotle's "Why do projectiles keep moving?", although addressing a very important issue, was actually the wrong question. The question should be "Why do projectiles *stop* moving?" Galileo realized that the natural thing for a horizontally moving object to do is to keep moving with uniform motion.

He then seized on this result to explain how the Earth could behave like other planets without requiring an engine to drive it (a problem we first addressed in Chapter 2). Galileo reasoned that horizontal was parallel to the surface of the Earth, and therefore circular. Thus it was natural for objects to travel on a circular path, without requiring an engine, unless there was some resistance due to a medium. But if the Earth were traveling in "empty space" (a vacuum), then there would be no resistance and the Earth could travel in a circular orbit forever, just as Copernicus had postulated. The Earth's atmosphere would also travel along with

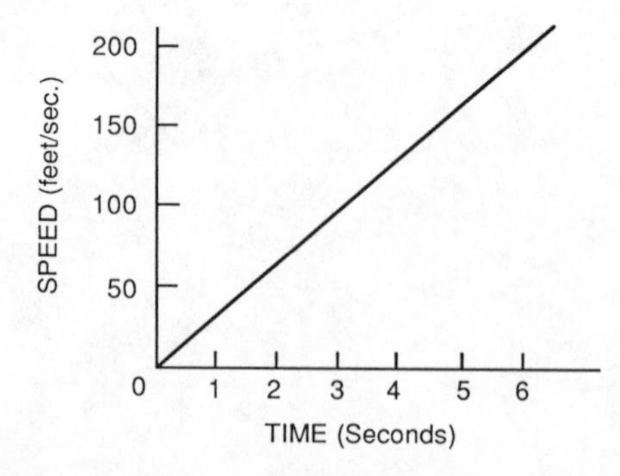

Figure 3-3. Results of Galileo's inclined-plane experiment shown on a speed-versus-time plot.

it, and thus there would also be no problem of high winds as a consequence of the Earth's motion. Unfortunately, Galileo had drawn the wrong conclusion from this aspect of his rolling-ball experiments. The ball rolling on the flat plane was truly rolling in a straight line, not on a level surface parallel to the Earth's surface.*

We now call the property of a body by which it keeps moving *inertia*. Perhaps the most dramatic example of inertia is the motion of a spacecraft in "outer" space. Recall how the astronauts traveled to the Moon. First, their spacecraft blasted off from Cape Canaveral and put them into orbit around the Earth. Next, after determining that everything was in order, they fired their rocket engines for several minutes in order to leave their Earth orbit and start toward the Moon. Once they reached the proper speed and direction, they shut off the engines. From there they were able simply to "coast" for about three days and 240,000 miles to a rendezvous with the Moon. Sometimes (but not always), they needed to fire their engines for a few minutes at about the halfway point, not to change their speed but to apply a "midcourse correction" to their direction. The point here is that the trip used the inertia of the spacecraft to keep it moving at a large constant speed (about 3000 miles per hour) all the way from the Earth to the Moon. It worked so well because essentially no friction resistance acted upon the craft in space. (Actually, the spacecraft slowed down somewhat until the Moon's gravitational attraction exceeded that of the Earth and then speeded up somewhat.)

Thus Galileo answered Aristotle's question by pointing out that it is natural for a moving object to keep moving. Of course, Galileo did not address why this is the natural state. But it was extremely important to first determine what nature does before it was possible to address questions of why things happen the way they do.

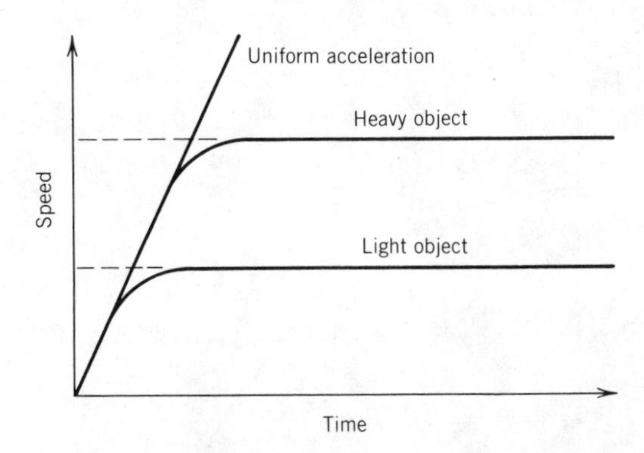

Figure 3-4. Graphical representation of motion of light and heavy falling objects.

*This misunderstanding could have been the reason Galileo failed to recognize the significance of Kepler's elliptical orbits. Galileo believed in circular orbits and was certain that his experiments had proved their validity—it would be difficult to explain a noncircular orbit. Not until the time of Isaac Newton was the proper explanation given of the physics of planetary orbits.

Thus Galileo (just as he did for falling objects) had for the first time properly recognized what happens in horizontal motion. In fact, Galileo went on specifically to consider projectile motion, the example that so perplexed Aristotle.

Galileo realized that the confusing thing about projectile motion (of thrown or struck objects) was that both horizontal and vertical motions were involved. He had already determined that "pure" horizontal motion was uniform motion in a straight line. But how should these be combined to describe a projectile? Galileo was also faced with this question in his use of inclined planes and pendulums to study falling. Both of these systems involve combined horizontal and vertical motions. Because he found that balls rolling down inclined planes always exhibited uniformly accelerated motion, no matter what the slope of the plane (although the magnitude of the acceleration increases as the slope increases), Galileo guessed that the effects of falling were independent of the horizontal motion. Because he found that all of his experiments with both inclined planes and pendulums gave results consistent with this hypothesis, he eventually concluded that it was always true. Galileo thus was led to what is known as the superposition principle, which states that for objects with combined vertical and horizontal motions the two motions can be analyzed separately and then combined to yield the net result.

According to the superposition principle, projectile motion involves both horizontal motion and vertical motion, which can be analyzed separately. Thus, the vertical motion is just falling motion—the same falling motion one would expect from an object with no horizontal motion. Similarly, the horizontal motion is uniform motion with whatever horizontal velocity the object has initially. The resulting combined motion for an object originally thrown horizontally from some height is represented in Figure 3-5.

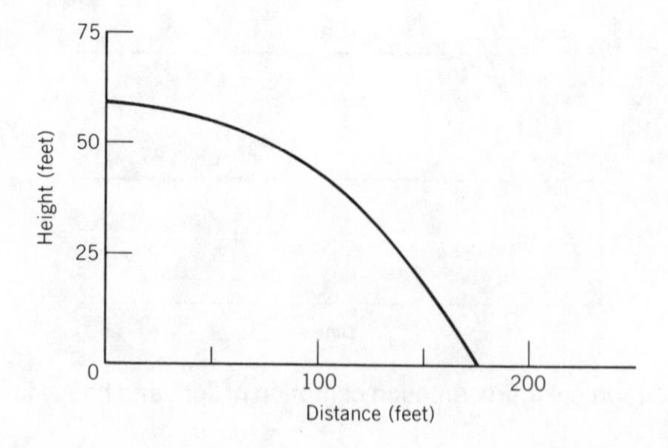

Figure 3-5. Graphical representation of height versus distance for an object thrown horizontally from a height.

The thrown object will strike the ground in the same amount of time as an object simply dropped from the same height. The thrown object will move horizontally at its initial horizontal speed until it strikes the ground. Thus, for example, if the horizontal speed of the object is 60 mph (88 feet per second) and it takes two seconds to fall to the ground (which would be the case if it starts from a height of 64 feet), the object will travel 176 feet horizontally (88 ft/sec $\times$ 2 sec). Note that the superposition principle says that a bullet shot horizontally from a high-powered rifle and a bullet dropped from the same initial height will hit the ground at the same time. The bullet fired from the rifle simply has a very large horizontal velocity and is able to move a rather large horizontal distance in the short amount of time available. (This result clearly does not apply to an object that "flies" in the air or in any way derives some "lift" from its horizontal motion. It is strictly valid only in a vacuum.)

Galileo also applied the concept of projectile motion to an apple (for example) falling from a tree fixed to the moving surface of the Earth. The apple maintains its "forward" motion while falling and thus does not get left behind, as had been claimed by some of the opponents of the heliocentric theory.

The net result of all of Galileo's contributions to the development of the science of mechanics is great. He determined exactly how objects fall and introduced the idea of inertia for moving objects. He was able to correct Aristotle's conceptual analysis of falling bodies and was able to answer his question of why projectiles keep moving. However, Galileo did not give the whys of the motions. As we will see, it was Isaac Newton who provided the "why" for falling motion.

Logic, Mathematics, and Science

Before discussing Newton's contributions to the science of mechanics, it is necessary to discuss some of the procedures and mental tools of science. In particular it is necessary to distinguish between inductive and deductive logic and to point out the value and use of mathematics in science. It is important to understand basically what is involved in the branches of mathematics known as analytic geometry and calculus, which will provide a better appreciation of Newton's contributions and how he made them. (These discussions will be useful also for the material in subsequent chapters.)

Two principal kinds of logic are used in science. When one performs, or collects the results from, a series of experiments and extracts a general rule, the logic is referred to as *inductive*. For example, if it is concluded from several carefully performed surveys that people prefer to watch mystery shows above all others, inductive logic is being used. The general rule was obtained directly from experiment and not through any line of reasoning starting from certain basic principles.

The famous English philosopher Francis Bacon (1561–1626) was an extreme proponent of the inductive method. He believed that if one were to carefully list and categorize all available facts on a certain subject, the natural laws governing that subject would become obvious to any serious student. While it is certainly true that Bacon made several important contributions to science in this manner in some of his studies and experiments, it is also true that he did not accomplish one of his stated goals—to rearrange the entire system of human knowledge by application of the power of inductive reasoning. He relied too heavily on this one kind of reasoning.

The other principle type of logic is *deductive*. In this style of logic, one proceeds from a few basic principles or theories and logically argues that certain other results must follow. Thus, if one theorizes that all people love to read a good mystery and argues that therefore mysteries will be popular television shows, one is using deductive logic. Perhaps the strongest proponent of the deductive method was the French philosopher Rene´ Descartes (1596–1650). He believed that one could not entirely trust observations (including experiments) to reveal the truths of the universe. Descartes believed that one should start from only a very few irrefutable principles and argue logically from them to determine the nature of the universe. Although Descartes made many important contributions to philosophy and mathematics (he developed much of the subject of analytic geometry discussed below), because of his insistence on starting with a few basic principles not obtained from observation he was not able, as he wished, to obtain a consistent description of the physical universe.

Isaac Newton skillfully combined the inductive and deductive methods. He started with certain specific observations and generalized from these observations to arrive at a theory indicating the general physical law that could explain the phenomenon. Once he arrived at a theory, he would deduce consequences from it to predict new phenomena that could be tested against observation. If the predictions were not upheld, he would try to modify the theory until it was able to provide successful descriptions of all appropriate phenomena. His theories always started from observations (the inductive method) and were tested by predicting new phenomena (the deductive method). Newton properly saw both methods as necessary in determining the physical laws of nature.

Before discussing some of the basic mathematical concepts that were developed about the time of Galileo and Newton (some by Newton himself) and that were necessary for Newton's synthesis of the science of mechanics, it is appropriate to discuss in a general way why mathematics is often used and even needed to describe physical laws.

Mathematics may be said to be the science of order and relationship. Thus, to the extent that the physical universe is orderly and has related parts, mathematics can be applied to its study. It is perhaps somewhat surprising that the universe is ordered and has related parts: the universe might not have been ordered and may have had no related parts, although such a universe would be hard to imagine. Nevertheless, it is still amazing that the more we learn about the universe, the more we know that it is ordered and appears to follow certain rules or laws. (As Einstein said, "The most incomprehensible thing about the universe is that it is comprehensible.") The more ordered and related the universe is found to be, the more mathematics will apply to its description.

Developments in physics often have followed developments in mathematics. Occasionally some areas of mathematics owe their origin to the need to solve a particular problem or class of problems in physics. We have already noted how Kepler was able to deduce the correct shape of the orbits of the planets because of his knowledge of geometrical figures, including the ellipse. He also made extensive use of logarithms, which had just been invented, in his many calculations. Similarly, Galileo needed proper definitions of motion and certain simple graphical techniques to determine the true nature of falling. We will see increased dependence on mathematics as we continue, and must now introduce two new mathematical developments in order to discuss Newton's works.

To use mathematics to describe physical motions and shapes it is necessary to find ways to combine algebra and geometry. The description of geometrical figures with mathematical formulas (and vice versa) is known as *analytic geometry*, a field largely developed just before the time of Newton.

Most simple and many complex figures can be described by mathematical formulas. Figure 3-6 provides several examples. The circle is seen to be described by a rather simple formula, as is the ellipse. Some of the more complicated patterns involve the use of trigonometric functions (sine, cosine, and the like) in their algebraic descriptions. It is not necessary here for us to understand how these algebraic descriptions (formulas) are obtained or even how they work. What is important is to be aware that such algebraic descriptions of figures often exist and are known.

In our discussion of types of motion and Galileo's studies of falling objects we sometimes found it necessary to describe the motion on a speed-versus-time graph and sometimes on a distance-versus-time graph. We indicated how some of these graphs were related; for example, how uniform motion would appear on both kinds of graphs. In general, the mathematical techniques for relating the features of these two kinds of graphs are part of the subject known as calculus. For example, the area under a speed-versus-time curve corresponds to the total distance traveled.

The subject of *integral calculus* deals with mathematical techniques for finding the area under any curve for which a mathematical formula exists (analytic geometry). Similarly, the slope of a distance-versus-time graph indicates the speed of an object and the slope of a speed-versus-time graph gives the accelera-

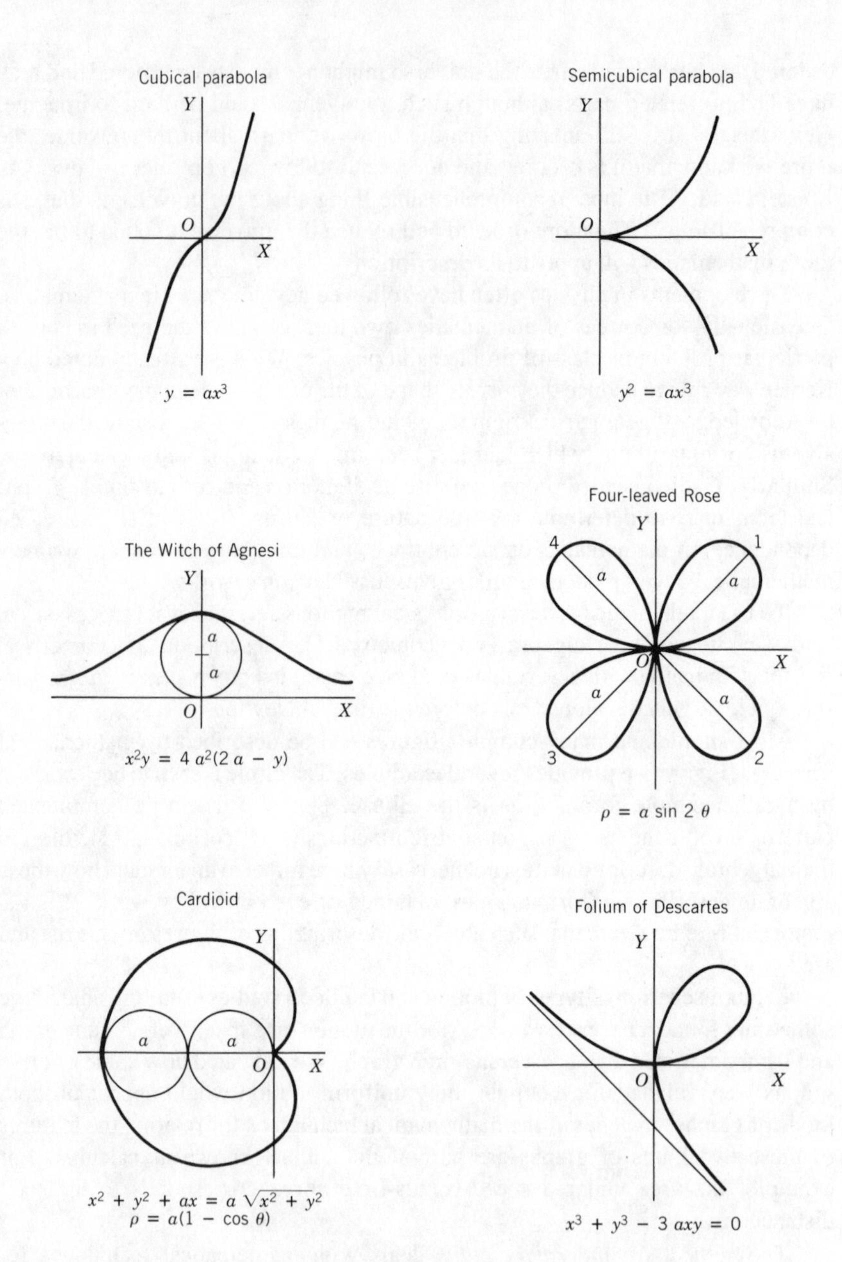

Figure 3-6. Various mathematical formulas and their graphical representations. (By permission from W. A. Granville, P. F. Smith, and W. R. Longley, *Elements of the Differential and Integral Calculus*. New York: John Wiley & Sons, 1962)

tion. The mathematical techniques for determining slopes of curves described by mathematical formulas are part of the subject matter of *differential calculus*.

Isaac Newton and the German mathematician-philosopher Gottfried Leibnitz (1646–1716) are credited as independent inventors of calculus. They performed their work separately but arrived at the same basic results on differential calculus. That Newton developed the ideas and methods of calculus certainly enabled him to arrive at some of his conclusions regarding mechanics and the universal law of gravity. He used calculus not only to translate from speed-versus-time graphs to distance-versus-time graphs, and vice versa, but also to find areas, volumes, and masses of figures and solids.

Finally, we must recognize the fact that some quantities have direction as well as magnitude (size). For example, if we say that an automobile is traveling at 30 mph we have not fully described its motion. We need to state also in what direction the car is traveling. Quantities that logically require both a direction and a magnitude are called *vector quantities*. The vector quantity associated with motion is called velocity. If we state the magnitude of the motion only, we refer to speed. Thus if we say that a car is traveling 30 mph in a northerly direction, we have specified its velocity.

Velocity is not the only quantity that is a vector; force is another example. Whenever a force is impressed on an object, it has both a certain strength and a specific direction that must be indicated in order to describe it completely. Similarly, acceleration (positive or negative) is always at a certain rate and in a specific direction, so that it also is a vector quantity. Thus circular motion, even with constant speed, is accelerated motion because its direction is always changing.

Some quantities do not require a direction but only a magnitude. Such quantities are called *scalar quantities*. The mass of an object (in kilograms, for example) needs no direction and is only a scalar quantity. The length of an object is also a scalar quantity. Other examples include electric charge, temperature, and voltage. Speed is a scalar quantity, but velocity is a vector. Because a proper discussion of motion, including the actions of impressed forces, must include directions, it is appropriate to use vector quantities. As we will see shortly, Newton was well aware of the need to use vector quantities in his descriptions of motion and made a special point to introduce them carefully.

Because vector quantities have both magnitude and direction, they can be represented graphically by arrows, as in Figure 3-7. The length of the arrow is proportional to the magnitude of the vector, and the orientation of the arrow and position of the arrowhead represent the direction. The simple mathematical operations carried out with ordinary numbers are only somewhat more complicated when carried out with vector quantities—including addition, subtraction, and multiplication. For the purposes of this and subsequent chapters, it is sufficient to describe graphically how vectors are added. To add two or more vectors,

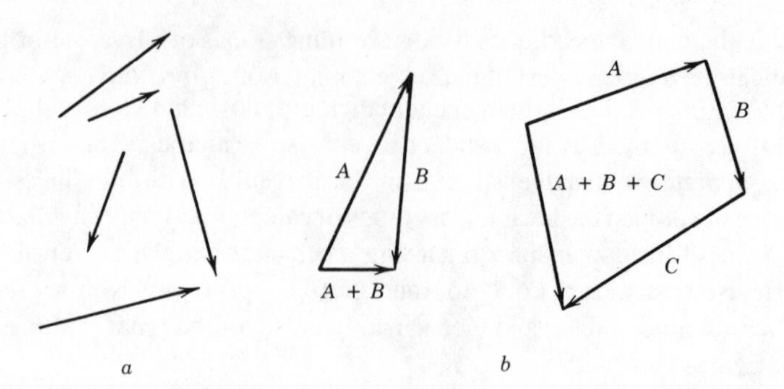

Figure 3-7. Vectors (a) Examples of vectors represented by arrows. (b) Examples of vector addition in graphical representation.

they are simply placed head to tail, maintaining their orientation and direction; and a final arrow or resultant vector is drawn from the tail of the first arrow to the head of the last arrow. This final arrow represents the vector sum of the individual vectors, as shown in Figure 3-7b. (One interesting result of vector addition is that a vector sum can have a smaller magnitude than each of the magnitudes of the vectors contributing to the sum.)

Newtonian Mechanics

We are now ready to discuss the contributions of Isaac Newton (1642–1727) to the science of mechanics. His work presents fairly complete answers to the questions first raised by Aristotle that have been the subject of this chapter. In fact, Newton's work represents one of the greatest contributions made by an individual to human understanding of the physical universe, and it would be difficult to overestimate the impact of his work on Western thought. He developed a picture of the universe as a subtle, elaborate clockwork slowly unwinding according to well-defined rules.

Isaac Newton was born in 1642, about a year after Galileo died. Newton continued and extended Galileo's work in several important ways, especially with regard to mechanics. Like Galileo, Newton was the dominant scientist of his generation. He made important contributions in mathematics, optics, wave phenomena, mechanics, and astronomy. Any of his several most important contributions would have established his place in history.

Newton became interested in experimental devices as a child and demonstrated an ability to make original designs of windmills, water clocks, and sundials. Because of his great potential, he was able to attend Cambridge University, where he showed exceptional mathematical ability. Following his graduation in 1665, Newton returned to his boyhood home at Woolsthorpe in Lincolnshire, where he lived with his widowed mother. This was the time of the Great Plague throughout Europe (some 31,000 people in London alone perished from the Black Death over a two-year period) as well as of a great fire that devastated much of London. Newton spent 1665 to 1667 in virtual seclusion at Woolsthorpe. It is clear now that during those two years he formulated the well-known binomial theorem of mathematics, developed calculus, studied the decomposition of white light into its spectrum (using a prism), and began his study of mechanics, including the universal law of gravitation. None of this work was published immediately, and some of it not for thirty years, but these were surely two of the most productive years ever spent by a single scientist.

In 1667 Newton returned to Cambridge as a lecturer. Newton's master at Cambridge, a noted mathematician named Isaac Barrow, became so impressed with Newton's work and abilities that he resigned his own chair of mathematics (in 1669) so that it might go to Newton. Newton was thus put into a position to follow his various studies freely. Unfortunately, Newton's first publications in the area of optics were not well received by other English scientists. Newton quickly became disillusioned and was reluctant to publish at all. He retreated into studies of alchemy, theology, and biblical prophecy (Newton was an ardent Christian).

Later, following some important suggestions from Robert Hooke and at the urging of an astronomer friend, Edmund Halley (for whom the famous comet was named), Newton finished the work on mechanics he had begun nearly twenty years earlier. In 1687 he published his most important work, *Principia Mathematica Philosophia Naturalis* (in English *The Mathematical Principles of Natural Philosophy*), or simply *Principia*—written in Latin, as were all scientific works at that time. The *Principia* represents one of the greatest accomplishments of the human mind and perhaps the single most important work in the history of physics.

Certainly Newton's work was the continuation of the earlier work of Aristotle, Kepler, and especially Galileo. He also obtained important ideas from Bacon, Descartes, and Hooke. Newton himself said "If I have been able to see so far, it is because I have stood on the shoulders of giants." Nevertheless, his work was more than the great synthesis of all the thinking which had been accomplished earlier, and as a result of the publication of the *Principia*, Newton was established as the outstanding natural philosopher of his time. He later served in Parliament (as the representative of Cambridge University) and became Warden of the Mint in London. Eventually Newton completely overshadowed all of his early denigrators and even became president of the prestigious Royal Society, with unparalleled power in the realm of science.

As a person, Isaac Newton was jealous, egotistical, complex, and troubled. An absent-minded confirmed bachelor, he was cared for by his niece. He was admired by Voltaire and Alexander Pope, despised by Jonathan Swift and William Blake. He made some of the most important contributions to science ever and his work ushered in a new wave of optimism regarding the ability of man to understand his universe.

Newton's *Principia* was written in a rigorously logical, axiomatic style, following the example set by the ancient Greek mathematician Euclid in his famous book on geometry. Newton presented a number of basic definitions and assumptions, defined what he called "fundamental" and "derived" quantities, and then presented his three laws of motion and a few other laws, including his universal law of gravitation, all based on induction from experiment. Then by logical deduction he showed that all the observed motions of objects in the physical universe, including celestial motion, natural local motion, and projectile motion, were simply consequences of his definitions and laws of motion. The *Principia* stands as a simple but truly monumental exposition of the nature of motion. Only in modern times, with the development of relativity and quantum mechanics, have any limitations on Newton's system been substantiated. These limitations are only important for systems involving extreme conditions: high speeds, small size, or very low temperatures.

Newton's definitions begin with the quantity of matter, which he described as the product of the density and volume of an object. This quantity describes "how much stuff" there is in an object and is now called *mass*. The quantity of motion is defined as the product of the quantity of matter of an object and its velocity—that is, its mass times its velocity (*mv*)—and is a vector quantity now called *momentum*. Both mass and velocity contribute to momentum. A 10-ton truck traveling at 20 mph has a greater magnitude of momentum than a subcompact automobile traveling at 50 mph, because the truck is about ten times more massive than the automobile, whereas the automobile has a speed only two and a half times greater than that of the truck.

Inertia is introduced as an inherent property of mass that describes its resistance to a change of its state of uniform motion in a straight line (discussed more fully below). Finally, Newton defined *impressed force* as an action that may change the state of motion (the momentum) of a body. Normally, one thinks of an impressed force as likely to speed or slow an object. Newton recognized that it is also possible for a force to change the direction of motion of an object without changing its speed. Such a force is called a centripetal force. An example is an object fastened to a string spun in a circle. The object may swing around with a constant speed, but is direction is continually changing. This change of direction is a form of acceleration and is caused by the impressed force from the string.

Newton next introduced his "fundamental quantities," which were required to be measurable and objective—that is, independent of the state of mind of the individual observer. Moreover, he desired that there be only a small number of these quantities so that the complete description of science would be based on as few basic ideas as possible. He needed only three such fundamental quantities: time, length, and mass. These were to be measured in terms of fundamental units. The modern scientific units are the second, the meter (39.37 in.), and the kilogram (equivalent to 2.2 lbs), respectively. Studies of electricity and heat have necessitated the introduction of two additional fundamental quantities: electric charge (the coulomb) and temperature (the Kelvin).

All other quantities can be expressed in terms of the basic quantities; for example, speed is expressed as a ratio of length to time (miles per hour). The magnitude of momentum is the product of mass and speed (mass times the ratio of length to time). Other quantities become more complex but can always be expressed in terms of the fundamental quantities. For example, kinetic energy is half the product of momentum and velocity squared (hence its units are mass times the square of the ratio of length to time).

Having defined various concepts and quantities carefully, Newton was able to introduce his three laws of motion very simply. These laws were intended to specify the relationship between impressed forces and the changes in the motions of an object.

> ***I. First Law of Motion—Law of Inertia***
> *In the absence of a net external force, an object will continue in a state of uniform motion (including rest) in a straight line.*

As already indicated, this law had been recognized by Galileo before Newton was born and represented a recasting of Aristotle's question "Why do objects keep moving?" into the form "Why do objects stop moving?" This law simply states that all objects with mass have a common property called inertia, which "keeps them doing what they have been doing." Newton, however, stated the law correctly in recognizing that inertial motion is straight-line, not circular, motion.

> ***II. Second Law of Motion—Law of Acceleration***
> *The time rate of change of motion (momentum) of an object is directly proportional to the magnitude of the impressed force and is in the direction of the impressed force.*

This law relates acceleration of an object to the impressed force. Note that momentum is mass times velocity, so if the mass of an object is not changing, then change of momentum implies a change in the velocity, which is called acceleration. Hence this law says that the acceleration of an object (with fixed mass) is proportional to the impressed force. Doubling the force doubles the acceleration, and tripling the force triples the acceleration. Note that both the impressed force and the consequent acceleration are vector quantities, with specific directions, which, according to the law, must be in the same direction. Newton's second law serves to tell us how the acceleration of an object depends on both the impressed force and the mass of the object. The dependence on the mass enters because the law states that it is the time rate of change of the *momentum* that is directly proportional to the impressed force, and momentum is the product of mass times velocity. Hence if the mass of an object is large, even a small change in velocity will produce a large change in the momentum. We see that for a given impressed force, which will produce a certain change in the momentum, the acceleration will be larger for an object with a small mass than for an object with a large mass. Formally, we say that the acceleration of an object is directly proportional to the net impressed force and inversely proportional to the mass of the object.

It is crucial to understand what "inversely proportional to" means. If A is inversely proportional to B, it means that if B gets larger, A gets smaller, and vice versa. Note that if A were directly proportional to B, if B became larger A would also get larger. An example of "directly proportional to" is the change in length of a stretched spring that increases as the stretching force increases. An example of "inversely proportional to" is the volume of a balloon full of air that decreases as the pressure of the surrounding atmosphere increases.

Thus Newton's second law tells us that the acceleration of an object is directly proportional to the impressed force and inversely proportional to the mass of the object. This is reasonable, of course, for we know that if we push or pull harder on an object, it will speed up faster, and that if the object is heavier (more massive), the same amount of push or pull will not be as effective. Larger engines give greater accelerations, but more massive cars reduce the acceleration. But Newton's second law goes beyond just what we know must be true in a qualitative way and tells us how acceleration is related to impressed force and mass in a quantitative way. The law indicates that the acceleration is not directly proportional to the square root of the force, for example, or the cube, or whatever, but simply is proportional to the magnitude of the force (that is, to the first power). Similarly, the second law indicates that the acceleration is inversely proportional to the mass in a specific quantitative way. Newton's second law gives one the ability to calculate what the acceleration of an object will be for a given force if the mass is known. Scientists and engineers use this formula possibly more than any other in their calculations. It applies to cars, spaceships, rockets, rubber balls, and subatomic particles. It is probably the most frequently used single equation in physics.

> ***III. Third Law of Motion—Law of Action and Reaction***
> *If one object exerts a force on a second object, the second object exerts an equal and oppositely directed force on the first object.*

Implicit in this law is the idea that forces represent the interaction between bodies. Whereas the first and second laws focus attention on the motion of an individual body, the third law says that forces exist because there are other bodies present. Whenever a force is exerted on one object or body by a second object, the first object exerts an equal and opposite force on the second object. The force relationship is symmetric.

For example, when a horse pulls a wagon forward along a road, the wagon simultaneously pulls back on the horse, and the horse feels that reaction force. The horse's hooves push tangentially against the surface of the road, if it is not too slippery, and simultaneously the surface of the road pushes tangentially against the bottom of the horse's hooves. It is important to remember which forces are acting on the horse. They are the reaction force of the road surface on the horse's hooves, which "pushes" forward, and the reaction force of the wagon as transmitted through the harness to the horse's shoulders, which "pulls" backward. In order for the horse to accelerate, the force of the road surface on the horse's hooves must be greater than the force of the harness on the horse's shoulders. But the force of the road surface is a reaction force and must be the same magnitude as the force exerted by the horse's hooves on the road. If the road is slippery there will not be enough friction between the road surface and the bottom of the horse's hooves, and the horse will slip. The horse will be unable to use its powerful muscles to push sufficiently hard against the road surface to generate a reaction force that will push the horse forward. These various forces which are all involved in the horse pulling a wagon are shown by arrows in Figure 3-8. It is essential to recognize, strange as it may seem, that it is the reaction force of the road surface against the horse's hooves that propels the horse forward.

A further example of the role of reaction forces and the third law of motion is supplied by a rocket engine. In a rocket, high-speed gases are expelled from the exhaust of a combustion chamber. The rocket exerts a force on the gases to drive them backward. Simultaneously the gases exert a reaction force on the rocket to drive it forward and this reaction force accelerates the mass of the rocket. (The corporate name of one of the major American manufacturers of rocket engines in the 1950s was Reaction Motors, Inc.)

Figure 3-8. Force pairs for a horse pulling a wagon. (a) Horse pulls on wagon (through the harness). (b) Wagon pulls back on horse. (c) Horse pushes against road surface. (d) Road surface pushes back against horse's hooves.

IV. Law of Conservation of Mass.

In addition to his three laws of motion, Newton considered it important to indicate two conservation laws he concluded must be true. In physics, a quantity is said to be conserved if the total amount is fixed (always the same). Newton's simplest conservation law is that for mass (quantity of matter). He believed that the toal amount of mass in a carefully controlled closed system is fixed ("closed system" simply means that nothing is allowed to enter or leave the system being studied). Certainly mass is not conserved in a system if more material is being added. The real import of this law is that Newton believed that mass could be neither created nor destroyed. He recognized that matter can change in form, as from a solid to a gas in burning. But Newton believed that if one carefully collected all the smoke and ashes from the burning of an object, one would find the same total amount of matter as existed in the beginning. We now know that this conservation law is imperceptibly violated in chemical reactions (such as burning) and strongly violated in nuclear reactions. These violations of Newton's law arise from the famous mass-energy relationship of Einstein's relativity, which we will discuss later.

V. Law of Conservation of Momentum.

Using his second and third laws of motion, Newton was able to show that whenever two bodies interact with each other, their total momentum (the sum of the momentum of the first body and the momentum of the second body) is always the same, even though their individual momenta will change as a result of the forces they exert on each other. Thus if two objects (an automobile and a truck, for example), collide with each other, their paths, speeds, and directions immediately after the collision will be different than they were before the collision. Nevertheless, if the momentum of the first object before the collision is added to the momentum of the second object before the collision and compared to the sum of the momentum of the first object and the momentum of the second object after the collision, the results will be the same. This is called the law of conservation of momentum, and it can be generalized to apply to a large number of objects that interact only with each other.

Although Newton derived the law of conservation of momentum as a consequence of his laws of motion, it is possible to use this law as a postulate and derive Newton's laws from it. In fact, the modern theories of relativity and of quantum mechanics show Newton's laws to be only approximately correct, particularly at high speeds and for subatomic particles, but the law of conservation of momentum is believed to be always exact.

As we have discussed, Galileo showed that if the effects of buoyancy, friction, and air resistance are eliminated, all objects fall to the surface of the earth with exactly the same acceleration, regardless of their mass, size, or shape. He showed also that even projectiles (objects that are thrown forward) also fall while they are moving forward. Their forward, or horizontal, motion is governed by the law of inertia, but their simultaneous falling motion is with exactly the same acceleration as any other falling object. It was left to Newton to analyze the nature of the force causing the acceleration of falling objects.

Newton considered that falling objects accelerate toward the Earth because the Earth exerts an attractive force on them. It is said that the basic idea for the law of gravity occurred to Newton when an apple fell on his head. This story is almost certainly untrue, but Newton did remark that it was while thinking about how gravity reaches up to pull down the highest apple in a tree that he first began to wonder just how far up gravity extended. He knew the Earth's gravity still existed high on mountains and concluded it likely had an effect in space. Finally, he wondered if gravity reached all the way up to the Moon and if one could, in fact, demonstrate that gravity was responsible for keeping the Moon in its orbit. He recognized that the Moon is in effect also "falling" toward the Earth, because its velocity is always changing, with the acceleration vector perpendicular to its circular path and therefore pointing to the Earth at the center of the Moon's orbit. This force, because it changes the direction of the Moon, is a *centripetal force*, discussed earlier. Similarly, the planets are "falling" toward the Sun because in their curvilinear elliptical paths their acceleration vectors point toward the Sun. Because accelerations can exist only when there are impressed forces, the Sun must exert a force on the planets. Therefore Newton concluded that the same kind of force, the force of gravity, might be acting throughout the universe.

By the use of Newton's laws of motion, and experimental measurements of the orbits of the Moon and the planets it is possible to reason by induction (infer) the exact form of the gravitational-force law. According to Newton's second law, if the force were constant, more massive objects should fall to Earth with a smaller acceleration than less massive objects. Therefore the Earth must exert a greater force on more massive objects—they weigh more (and weight is a force). In fact, the gravitational force on an object must be exactly proportional to the the mass of the object in order that all objects have the same acceleration. Therefore the Earth exerts a force on an object proportional to the mass of the object. But according to Newton's third law, the object must exert an equal and opposite force on the Earth and this force must be proportional to the mass of the Earth, because from the object's point of view the Earth is just another object. Both proportionalities must apply, and therefore the force of gravitation must depend on both the mass of the Earth and the mass of the object.

Newton knew that the acceleration of the Moon while "falling" toward the Earth (he considered the Moon a projectile) was much less than the acceleration of objects falling near the surface of the Earth. (Knowing the dimensions of the Moon's orbit and the length of time it takes the Moon to go around its orbit once—27 1/3 days—he was able to calculate the acceleration of the Moon.) He eventually concluded that the Earth's force of gravitation must depend on the distance between the center of the Earth and the center of the object (the Moon, in this case). The distance from the center of the Earth to the center of the Moon is 60 times greater than the distance from the center of the Earth to an object on the surface of the Earth. The calculated "falling" acceleration of the Moon is 3600 times smaller than the falling acceleration of an object near the surface of the Earth. But 3600 equals 60 times 60; therefore it is a good guess that the force of gravity decreases with the square of the distance from the center of the attracting object. Putting all these considerations together, Newton proposed his law of gravitation.

> ***VI. The Universal Law of Gravitation***
> *Every material particle in the universe attracts every other material particle with a force that is proportional to the product of the masses of the two particles, and inversely proportional to the square of the distance between their centers. The force is directed along a line joining their centers.*

In mathematical form, the magnitude of the force is represented by

$$F_{gravity} = Gm_1m_2/r^2$$

where $F_{gravity}$ is the magnitude of the force, m_1 is the mass of one of the objects, m_2 is the mass of the other object, r is the distance between the centers of the two objects, and G is a proportionality constant (which is needed just to use units already defined for mass, distance, and force).

The universal law of gravitation is referred to as an *inverse square law* because the magnitude of the force is inversely proportional to the square of the distance between the objects (see the discussion on Newton's second law above). Newton was not the first person to consider that the law of gravitation might be an inverse square law, of course. A few other European scientists had contemplated or were also considering such a dependence, especially the English physicist Robert Hooke and the Dutch physicist Christian Huygens. Because they lacked Newton's mathematical skills (especially the ideas of calculus), however, they were unable to demonstrate that an inverse square law works to describe falling objects and the motion of the Moon and of the planets.

With his three laws of motion and the universal law of gravitation, Newton could account for all types of motion: falling bodies on the surface of the Earth, projectiles, and the heavenly bodies in the sky. In particular, he was able to account for Kepler's laws of planetary motion. Newton was able to show that the three laws of planetary motion are logical mathematical consequences of his laws of motion and of his universal law of gravitation. For example, by using the mathematical form of the universal law of gravitation for the impressed force in his second law of motion, he was able to obtain the mathematical formula for an ellipse (using analytic geometry), thus explaining Kepler's first law of planetary motion. Similarly he was able to "explain" why a planet in an elliptical orbit moved with varying speeds (Kepler's second law) and how the period for a complete orbit depends on distance from the Sun (Kepler's third law).

Newton's law of gravitation applies not only to the Earth and the Moon but also to the Earth and the Sun, the other planets and the Sun, Jupiter and its moons, and so on. An analysis similar to that for Earth and Moon also works for these other examples. The law of gravitation as well as the laws of motion apply throughout the solar system. Moreover, astronomical evidence shows that they apply also to the universe on an even larger scale. These laws apparently are truly universal.

Newton's laws allow one to go beyond Kepler's laws of planetary motion. Because Newton's law of gravitation indicates that all material objects exert a gravitational force on all other bodies, we know that the planets should experience gravitational attraction not only from the Sun but also from each other. Thus the planets should disturb each other's orbits, so the resulting orbits are not perfectly elliptical. Newton was concerned about the effects of these perturbations. He thought that the planetary orbits might be unstable enough for the gravitational pulls of the other planets to cause a planet to leave its orbit. Newton speculated that divine intervention might be required to maintain the planets in their orbits. Nearly a century later, the famous French mathematician and astronomer Pierre Laplace (1749–1827) showed that the planetary orbits are actually stable, so that although the orbits are not elliptical the planets will not leave their orbits if perturbed.

Eventually the perturbations caused by the planets' gravitational attractions for each other provided some of the most convincing evidence for the correctness of Newton's laws. In 1781, a new planet was discovered and named Uranus. Using improved mathematical techniques to calculate the effects of all the other planets on the orbit of Uranus, by about 1820 it became clear that there were discrepancies between the observed orbit and the calculated orbit. Suspecting that there was another planet in the solar system, the French astronomer U. J. J. Leverrier and the English astronomer J. C. Adams eventually were able to calculate just where and how massive the new planet must be in order to explain the discrepancies. In 1846, Neptune was discovered exactly where they had predicted. Careful study of the orbit of the new planet Neptune indicated that yet another planet must exist. In 1930, Pluto was discovered and most of the remaining unexplained perturbations were removed. A debate continues about whether there are now any residual discrepancies in the orbits of all the planets, after taking the gravitational attractions of all the presently known ones into account. The remaining discrepancies are small, and they may be only slight errors in the difficult observations—or there may be yet another planet.

The important point of the planetary perturbations is that, in the end, they yielded a remarkable test of Newton's laws. The effects of all the planets on each other are predicted and observed, and the agreement is remarkable. Only in the case of the planet Mercury have the calculations been significantly off the mark. For Mercury Einstein's general theory of relativity is required to resolve the discrepancy (see Chapter 6).

In addition to planetary orbits, Newton's laws have been shown to be highly accurate in many other physical situations. Calculations similar to those for the planets are made for artificial satellite orbits around the Earth. The appropriate combinations of Newton's three laws of motion and the universal law of gravitation explain all projectile motions. Newton's laws are used daily by engineers and scientists to describe the motions of everything from trains to subatomic particles. Lawyers often study Newton's laws in order to be able to analyze automobile accidents. Although we will see that relativity and quantum mechanics will require some modification and extension of Newton's laws, they are continually being demonstrated to be accurate descriptions of motion in the physical universe.

It is important to realize how much Newton combined the use of inductive and deductive logic. He always started with a critical observation. In the case of gravity, for example, he began by analyzing how objects fall and then extended his analysis to the motion of the Moon. Once he had a general theory, arrived at from his observations by inductive logic, he then proceeded to use the general theory to explain or predict other phenomena by deductive logic. In the case of gravity, he showed that his inverse square law could explain Kepler's laws of planetary motion. Newton set a clear example, followed since by nearly all scientists, of combining inductive and deductive logic to obtain and test a theory.

Before we consider the philosophical consequences and implications of Newton's work, it is useful to point out that his laws do not answer all questions with regard to motion and gravity. Although Newton explained, using an inverse square law for gravity, why objects fall with uniformly accelerated motion, he did not indicate why gravity has such a dependence on distance, or how gravity acts over a distance. This latter question is known as the action-at-a-distance problem and has puzzled scientists for a very long time. Gravity can act across apparently empty space. The Earth attracts the Moon (and vice versa) and the Sun attracts the Earth (and vice versa). The Earth and the Moon are separated by about 240,000 miles of almost perfectly empty space; the Sun and the Earth are separated by about 93 million miles of similar space. It is quite remarkable that, somehow, these bodies interact with each other. A similar situation exists for magnetic forces. It can be quite intriguing to play with two small bar magnets. It is easy to feel them start to attract or repel each other before they actually touch. How do they do it? This is another version of the action-at-a-distance problem.

We are not ready yet to fully discuss the answer to this problem, which we will consider again in the chapter on relativity and then discuss in more detail in the final chapter, on the fundamental building blocks of nature. Briefly, it appears that bodies interact, whether by gravity or electromagnetic forces or other forces, by exchanging tiny subatomic particles invisible to the human eye. Newton discussed the action-at-a-distance problem but, unable to solve it, concluded by saying "I make no hypothesis."

Summarizing, we may say that Galileo showed exactly how objects fall: with uniformly accelerated motion. Newton then explained that objects fall this way because of certain laws of motion and the inverse square law of gravity. We now know how forces are transmitted, but we still do not know why objects have inertia or why gravity exists at all. We will make further progress on these questions as we discuss relativity, quantum mechanics, and elementary particles, but we will see that every answer to a question always seems to raise a new question.

Consequences and Implications

Newton's great synthesis of the laws of motion and the universal law of gravitation to explain all the kinds of motion had a great philosophical and emotional impact on the scientists of the time. His work created an image of a great clockworklike universe. Because it is only in relatively recent times that this image has been questioned, it is important to understand what this image was and some of the philosophical consequences it implied.

Newton's laws indicated that the motions and interactions of all the material bodies of the universe obey a few certain, relatively simple laws. According to these rules, if one knows the locations, masses, and velocities of a set of objects at some one instant, one can determine how the objects interact with each other and what will be the result of collisions. For example, one uses the law of gravitation to determine what net force is impressed on each object because of the presence of all the other objects. Next, one can use the second law of motion to determine the magnitude and direction of the consequent acceleration of that object, which allows one to know the velocity of the object. The law of conservation of momentum will enable one to determine the result of any possible collision. Now, of course, this can be quite involved, but it is often performed for certain sets of bodies by engineers and scientists. And it works!

The motions of a set of billiard balls on a table or of subatomic particles in nuclear reactions can be accurately predicted if one knows at some one time where each object is and what its velocity is. Forensic physicists can reliably calculate backward to determine the initial directions and speeds of automobiles involved in an accident if the final directions and speeds can be determined from skid marks and similar clues. Newton's laws indicated that the future motions of any set of objects can be determined if we know their initial motions.

If one takes this realization to its apparent logical conclusion, it implies that the future motions of all the objects of the universe could be predicted if we could only determine what they are right now. Of course, we cannot determine the locations and velocities of all the bodies of the universe at any one moment (not even with the world's largest computer). But philosophers after Newton were quick to realize that the important point was that in principle it could be done. Whatever the positions and velocities are now completely determines what they will be in the future. It does not matter that we cannot carry out the calculations of the future of the entire universe; it is still completely determined. Newton's laws indicate a universe evolving in time, according to his laws, in a completely predetermined fashion. The universe is like a giant, elaborate clockwork, operating in a prescribed way.

Many philosophers cited Newton's work as proof of predestination. If it is all evolving in a (theoretically) predictable way, it is predetermined. Newton's universe was evoked to challenge the view of self-determination of individuals —and even of divine election of kings. Such conclusions may appear valid. As we will see later with regard to the apparent consequences of other physical theories, however, such grand conclusions applied to the universe as a whole are questionable. Specifically, we will see that the results of quantum mechanics directly challenge this predetermined universe. More generally, it is hard to be sure of the extension of the specific laws of nature determined in our small region of the universe to the net course of the universe as a whole.

Newton's laws were a great triumph for science. So much of physics had appeared to be separate, unrelated phenomena with no simple explanations. Newton explained all kinds of motion in terms of only a few simple laws. Scientists were extremely encouraged. They became confident that our universe was indeed a rational system governed by simple laws. It appeared that perhaps there were good reasons for everything.

Scientists in all areas set out anew to discover the "basic" laws of their fields. In fact, areas not regarded previously as sciences were declared to be so. The "social sciences" were created and scholars started to search for the laws that govern human behavior. There was a movement to make a science of every field of human endeavor. Adam Smith in his *Wealth of Nations* tried to define the laws of economics so that scientifically sound economic policies could be adopted. Auguste Comte tried to do the same in sociology. Faith in the ultimate power of the "scientific mind" was tremendous. The kings of Europe established academies dedicated to solving all human problems. Frederick the Great and Catherine of Russia founded scientific academies in imitation of those of Charles II in England and Louis XIV in France. They spoke of the time as an Age of Reason and of Newton as a bearer of light. The point cannot be made too strongly: Newton's work unified a very diverse range of phenomena by explaining all motion with only a few laws. If the physical universe was so simple, the thinking proceeded, why not all branches of knowledge?

The idea that there are simple, logical rules for everything extended beyond mere scholarly pursuits. The American and French revolutions were both dedicated to the idea that people had certain natural rights which must be honored. The designers of the new governments proceeded to try to establish laws and governing principles consistent with these natural rights and tried to argue logically from the assumed rights to deduce the kind of government that should exist. The *Declaration of Independence* has the same kind of axiomatic structure, with deductive logic supporting the final conclusions, as does Newton's *Principia*. Benjamin Franklin, who was one of the best scientists of his time, wrote an essay entitled "On Liberty and Necessity; Man in the Newtonian Universe." Jefferson called himself a scientist and began the American tradition of governmental encouragement of the sciences. The Age of Reason was in full bloom.

James Joule

4 The Energy Concept

Energy is what makes it go

Although Newton's laws of mechanics were extremely successful in explaining many natural phenomena, there still remained many processes in nature that could not be understood solely by application of those laws. One of the best-known of these other phenomena was heat. Although various attempts had been made to understand heat from at least the time of Aristotle, most of our present knowledge of heat was developed in the nineteenth century. The studies performed during the nineteenth century in order better to understand heat and its production eventually led to the modern concept of energy. Both because this concept is most significant for understanding the nature of our physical universe and because energy has become so important a topic in our lives, it is worthwhile to retrace the important steps in the development of the energy concept. It is important to know what energy is, how it is characterized and measured, what forms it can have, and how it can be converted from one form to another.

Interactions and Conservation Laws

The first real step in the development of the energy concept was the realization that there are conservation laws in nature. If there is a conservation law for some quantity, then the total amount of that quantity is maintained at a constant value for all time in an isolated system. Newton developed the law of conservation of momentum in his *Principia*. Newton believed that the total momentum (he called it motion) of a system, which was characterized by the sum of the products of the mass of each object and its velocity, remained a constant, even if the various masses were allowed to collide with each other. An "isolated system" is a collection of objects that can interact with each other but not with anything else. As the objects interact, say by collisions, their individual velocities and therefore their momenta may change; however, the vector sum of all their momenta will always have the same value. (The idea for this conservation law was not original with Newton but had been developed by the English scientists Hooke, Wallis, and Wren and proposed as a law by the Dutch physicist Christian Huygens in 1668.)

For example, this conservation law allows one to predict what will happen when one object with a known amount of momentum (mass times velocity) collides with another object of known momentum. The momentum of each of the objects after the collision can be calculated if one other fact about the collision is known: for example, whether the two objects stick together or whether the collision is elastic. If we can discover more "laws" like this that nature always obeys, we can increase our ability to predict exactly what will happen in increasingly complex situations because every conservation law can be expressed as an equation. Because equations can be solved mathematically, exact results can be predicted.

Christian Huygens also recognized that in certain collisions another quantity besides momentum is conserved (that is, its total value remains constant). This quantity, called *vis viva* (living force) by Leibnitz, was calculated as the product of the mass of an object times the square of its velocity (mv^2). Some time later a factor of one-half was included and the quantity was renamed *kinetic energy*. (The factor of one-half follows when this quantity is derived from Newton's laws of motion.) The kind of collision in which kinetic energy is conserved is known as an elastic collision. Billiard-ball collisions are familiar examples of essentially elastic collisions.

The significance of the concept of kinetic energy of an object was enhanced when it was recognized to be the result of applying a force to the object. If one measures the amount of the net applied force and the distance over which the force is applied to the object, the gain in kinetic energy for the object will be exactly equal to the product of the force times the distance.

The force need not be constant but may vary in any manner over the measured distance. The kinetic energy acquired by the object is then equal to the average force times the total distance. An increase of kinetic energy is thus the integrated effect of an accelerating force acting over a distance.

In fact, the converse of this relationship shows why kinetic energy was first known as *vis viva*. If an object has velocity, it has the ability to exert a force over a distance; for example, a large moving ball colliding with a spring can compress the spring more than the "dead weight" of the ball alone. This ability to exert additional force is associated with the motion of the object and is measured by its kinetic energy. (Note that the kinetic energy of the object decreases as it compresses the spring.) This force due to motion alone was therefore called *vis viva*, or living force.

Similarly, the change in momentum of an object (mass times velocity) can be demonstrated to be the effect of a force on an object multiplied by the length of time the force is allowed to act. The momentum added to the object will be equal to the average force times the total time.

Both kinetic energy and momentum are related to the velocity of an object. Kinetic energy is a scalar quantity, whereas momentum is a vector quantity (Chapter 3). Both quantities are determined by the integrated (cumulative) effect of the action of net force. However, kinetic energy is the integrated effect of net force acting over a distance as contrasted with momentum, which is the integrated effect of a force acting over time.

The product of force times distance is a much more general concept than indicated so far. It is an important quantity, and is called work. One must be very careful when using the word *work,* however, because it means something very specific. A physicist would say that if you pushed very hard against a wagon full of bricks and the wagon did not move, you did no work on the wagon! Only if the wagon moved as a result of the applied force would one have done some work on it. One can double the amount of work done either by doubling the applied force or by doubling the distance over which the force is applied. In either case, the net effect will be the same—twice the amount of kinetic energy will have been imparted to the object.

Kinetic energy is our first example of one of the forms energy may have. Whenever we do work (in the physicist's sense) on an object, we change the energy of the object. One can work on an object and not change its kinetic energy, however. For example, the force acting over the distance may just be moving an object against gravity slowly up an inclined plane. When the force is removed, the object is left higher up the ramp but not moving. If work has been done on the object, what kind of energy has been imparted to it? The answer is *potential energy* because the object, in its higher position, now has the potential to acquire kinetic energy. If the object were nudged off the side of the ramp to make it fall, it would accelerate and gain kinetic energy as it fell. Potential energy is energy associated with an object because of its position or configuration, as discussed further below. (It was originally called latent *vis viva* because it was considered to have the potential of being converted into *vis viva*.)

The concept of potential energy makes it possible to find another conservation principle involving motion. There are certain common systems not involving collisions, but nevertheless involving movement, for which some aspect is unchanging. Something is being conserved, physicists prefer to say. One such system is a pendulum. A pendulum, if constructed properly, will continue to swing for a very long time. If we watch such a pendulum for only a short time we can analyze it as if it were not going to stop at all. More precisely, we can measure just how high the pendulum bob rises on every swing and see that it comes back to almost the same height every time. It would not be hard to convince ourselves that if we could just get rid of the effect of air resistance and put some extremely good lubricant on the pivot, the pendulum bob would always return to exactly the same height on every swing. This is an interesting observation. What is it that makes nature act this way for a pendulum?

Is there something being conserved in the pendulum's motion that requires it to behave this way? If so, what is being conserved? It is not kinetic energy because the pendulum bob speeds up and slows down, even coming to a stop momentarily at the top of each swing. When the bob passes through the bottom position, it is moving quite fast. Hence, because kinetic energy is one-half of mass times the square of the velocity ($1/2mv^2$), it is zero at the high points of a swing and a maximum at the low point.

Mathematical analysis shows that the quantity being conserved is the sum of the kinetic energy and the potential energy of the bob. The potential energy is said to be zero when the bob is at its lowest position and increases directly with the height above that position. It reaches its maximum when the bob is at the top of the swing where the kinetic energy has gone to zero. This works out just right. The total mechanical energy (the sum of these two kinds of energies) is always the same. The pendulum is an example of a conservative system.

Another example of a conservative system is a ball rolling up and down the sides of a valley. If the terrain is very smooth, and one starts a ball at a certain height on a hill on one side of the valley, when released it will roll down into the valley and up the hill on the other side to the same height it had on the first side. As it moves, it will possess the same combination of potential and kinetic energies as a pendulum bob. The sum, the total mechanical energy, will again always be a constant.

It can be seen why this new form of energy is called potential energy. When the ball is high on one hill, it has a great amount of "potential" to acquire kinetic energy. The higher up the hill the ball is started, the faster it will be moving, and therefore the more kinetic energy it will have when it reaches the valley. Note that this potential energy is actually gravitational potential energy. It is against the force of gravity that work is done to raise the ball higher, and the force of gravity will accelerate the ball back down into the valley.

Another system that also involves only motion and the gravitational force is the system of a planet moving around the Sun, also a conservative system. Kepler's first law of planetary motion (Chapter 2) states that the planet's orbit is an ellipse, with the Sun at one focus. Thus the planet is sometimes closer, then sometimes farther from the Sun. From Newton's universal law of gravitation it can be shown that when the planet is farther from the Sun, it has more gravitational potential energy (it has farther to "fall" into the Sun, so it is "higher"). Kepler's second law of planetary motion states that the radius vector from the Sun to a planet sweeps out equal areas in equal times. This law accounts for the known fact that the planet moves faster when it is closer to the Sun. Again, one can see that things occur in the right way: The planet moves faster when it is closer to the Sun, so the planet has more kinetic energy when its potential energy is less. Mathematical analysis again shows that the sum of the potential and kinetic energies will remain exactly constant. (Kepler's second law also illustrates another conservation principle—conservation of angular momentum, discussed in Chapter 8.)

In all these examples the motion has been discussed in terms of velocity and position. It was not necessary to discuss acceleration or Newton's laws of motion as applied to those cases. The principle of conservation of total mechanical energy permits us to analyze the motion very simply. This is one of the major reasons for using the conservation principle: it is easier. Of course, the proof of the principle is obtained from Newton's laws of motion, but they are assumptions anyway, even though based on observation. One could equally well start with the conservation principle as a basic assumption, thereby making physics "simpler."

Conservative systems thus appear to be operating in our universe. These systems always seem to maintain the same value for the total energy when taken as the sum of the kinetic energy plus the potential energy. But these systems are actually only approximately conservative. The pendulum will slowly come to a stop, as will the ball in the valley. The total energy appears to go slowly to zero. The planets are slowly moving out into larger orbits. Is there or isn't there an exact conservation law for these systems?

The answer to the preceding question is that the total energy actually is conserved, but there exists yet a third kind of energy, and the kinetic and potential energies are slowly being converted into this third form. The new form of energy to be considered is *heat*. The systems have frictional forces (between the pendulum bob and the air and at the pivot for the first example), and friction (as people who have rubbed their hands together to warm them know) generates heat. The heat so generated is gradually lost to the system. It escapes into the surrounding air and slightly warms it. That all of the lost energy appears as heat, and that heat is another form of energy, was established by nineteenth-century physicists. It is worthwhile to discuss now, in some detail, the evolution of our understanding of heat as a form of energy. In fact, historically, physicists seriously considered that there was a conservation law for total energy only after heat was shown to be a form of energy.

Heat and Motion

Heat has been studied from very early times. Aristotle considered fire one of the five basic elements of the universe. Heat was later recognized as something that flowed from hot objects to colder ones and came to be regarded as a type of fluid. About the time of Galileo this heat fluid was known as phlogiston and was considered to be the soul of matter. Phlogiston was believed to have mass and could be driven out of or absorbed by an object when burning.

In the latter part of the eighteenth century the idea that heat was a fluid was further refined by the French chemist Antoine Lavoisier and became known as the caloric theory of heat. The caloric theory was accepted by most scientists as the correct theory of heat by the beginning of the nineteenth century. The fluid of heat was called the caloric fluid and was supposed to be massless, colorless, and conserved in its total amount in the universe. (Lavoisier developed this theory while proving that mass is conserved when the chemical reaction known as oxidation, or burning, takes place.) The caloric fluid, with its special characteristics, could not actually be separated from objects in order to be observed or studied by itself. The caloric-fluid theory thus presented a growing abstractness in the explanations of heat phenomena. Several phenomena were accurately understood in terms of the caloric theory, even though it was itself incorrect.

As an example of a process that was thought to be understood by analysis with the caloric theory of heat, consider the operation of a steam engine. This is a good example both because improved understanding of the nature of heat came from attempts to improve steam engines and because it introduces several important concepts needed for the study of heat. The first working steam engine was constructed about 1712 by Thomas Newcomen, an English blacksmith. Very rapidly the Newcomen engine was installed as a power source for water pumps in coal mines throughout England. It replaced cumbersome and costly horse-team-powered pumps.

A simplified Newcomen steam engine is shown in Figure 4-1. The fire under the boiler continually produces steam from the water in the boiler. The operation of the engine is accomplished in essentially four steps performed by using the two valves labeled *steam valve* and *water valve*: (1) open the steam valve to let steam into the piston chamber, the pressure of the steam causing the piston to rise and the pump rod to descend; (2) close the steam valve; (3) open the water valve, which allows cool water to spray into the chamber, condensing the steam and creating a partial vacuum, thereby "sucking" the piston down and lifting the pump rod; and (4) close the water valve. The whole cycle is then repeated. In the earliest versions of the steam engine, the valves were actually opened and closed by hand. Later it was realized that the engine itself could "automatically" open and close the valves by the use of gears and levers.

The early ideas regarding the essentials of the steam engine were very crude by today's standards. Although it is called a steam engine, the fuel being burned under the boiler actually provides the power for the engine. Early experimenters were not entirely convinced of this, however. The power source for the steam engine was considered the steam and the efficiency of the engine was measured in terms of the amount of steam it consumed. Many of these early ideas did improve the steam engine considerably, especially those of the Scottish inventor James Watt, who patented the first really efficient steam engine in 1769.

In the Newcomen engine, the walls of the cylinder are cooled during the condensation or down-stroke step of the cycle. When steam is admitted into the cylinder for the expansion or up-stroke step of the cycle, about two-thirds of the steam is consumed just to reheat the walls of the cylinder so that the rest of the steam can stay hot enough to exert sufficient pressure to move the piston up. Watt realized that the condensation step was necessary for the cyclic operation of the engine, but he also realized that the actual cooling of the steam could take place in a location other than the hot cyclinder. He therefore introduced into his engine a separate chamber called the condenser to carry out the cooling for the condensa-

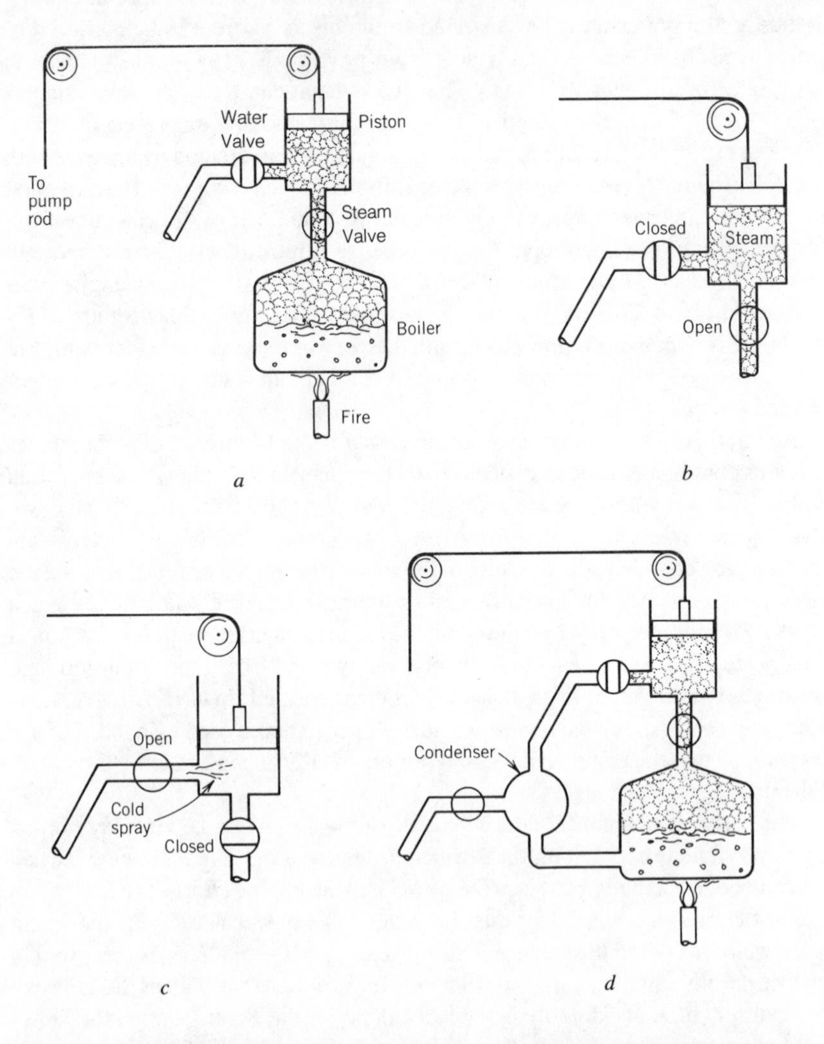

Figure 4-1. Schematic representation of early steam engines. (a) The Newcomen engine. (b) Steam pressure pushes piston up. (c) Condensing steam pulls piston down. (d) Watt's improved steam engine with separate condensing chamber.

tion step, as shown in Figure 4-1d. The condenser was placed between the water valve and the piston chamber, and between steps 2 and 3 above, the steam was sucked into the condenser by a pump (not shown in the diagram) and Step 3 was modified appropriately. Also, the water from the condenser was returned to the boiler.

Watt also recognized that the necessary condensation step resulted in the loss of some steam that was capable of pushing the piston further. To decrease the relative amount of steam lost in the condensation step, he decided to use higher-pressure (and, therefore, higher-temperature) steam in the initial stages of the expansion step. Thus, even after the steam-inlet valve was closed, the steam in the cylinder would still be at a pressure high enough to continue expanding and pushing the piston upward. While it continued expanding, its pressure would drop and it would "cool" but would still be doing work. Only after it had done this additional work would the condensation step begin.

With these and other improvements, Watt's steam engine was so efficient that he was able to give it away rather than sell it directly. All the user of the engine had to pay Watt was the money saved on fuel costs for the first three years of operation of the engine. Watt and his partner Matthew Boulton became wealthy, and the Industrial Revolution in England received a tremendous impetus from a new source of cheap power.

A major step in understanding the basic scientific principles underlying the operations of a steam engine was taken by a young French military engineer named Sadi Carnot. In 1824 he published a short book, *On the Motive Power of*

James Watt

Fire, in which he presented a penetrating theoretical analysis of the generation of motion by the use of heat. Carnot was considerably ahead of his time, and so his contribution went essentially unnoticed for about twenty-five years.

In considering Watt's work, Carnot realized that the real source of the power of the steam engine was the heat derived from the fuel and that the steam engine was simply a very effective means of using that heat to generate motion. He therefore decided to analyze the fundamental way in which heat could generate motion, believing that once this was understood it might be possible to design even more efficient engines than Watt had developed. He made use of the then-accepted caloric theory of heat and arrived at some remarkable conclusions despite an erroneous theory. In time, after the publication of his book, he realized that the caloric theory might be in error, but he died of cholera before he could pursue the matter much further. Carnot nevertheless explained the performance of a steam engine in terms of an analogy that is extremely useful for understanding the motive power for heat.

According to the caloric theory, heat is a fluid that flows from hot objects to colder ones. The steam engine is understood to be the heat equivalent of the mechanical waterwheel. The waterwheel derives its power from the flow of water from a higher elevation to a lower. The temperatures of the objects can be regarded as analogous to the elevations of the entrance stream and the exit stream for the waterwheel. It is commonly known that for a constant amount of water flowing over the waterwheel, the larger the wheel, the more work that can be done (in other words, the farther the water must fall). Hence, Carnot suggested that a steam engine (he said it should more properly be called a heat engine) could do more work with the same amount of heat (caloric fluid) if the heat was made to flow over a larger temperature difference.

One must be somewhat clever to utilize this suggestion. The temperature difference in a steam engine is between the heat of the steam and the temperature of the surroundings. The steam is at the boiling point of water. At first it appears that we cannot control either the high temperature or the low temperature. But Carnot pointed out that when the steam is produced at a higher pressure, the temperature of the steam is higher than the normal boiling point of water (the basis for the effectiveness of a pressure cooker). Thus higher pressure is actually a means of getting higher temperature and a larger temperature difference. Carnot thus explained the improvement Watt found by using higher pressures. All modern steam engines operate at high pressures and accordingly realize an increase in efficiency.

The caloric theory thus seemed to explain the steam engine very nicely. Some other aspects of heat flow also appeared to be accounted for by the caloric theory. Some concepts and definitions that will be needed again later can be introduced here in terms of the caloric theory.

One of the first things noticed about heat flow is that it is always from hot to cold. When two objects at different temperatures are placed in contact with each other, the colder one becomes warmer and the warmer one becomes colder. One never observes that the warmer one gets warmer still and the colder one colder. The caloric theory accounts for the observed fact that heat always flows from hot to cold by simply stating that temperature differences cause heat flow. Heat flows "down temperature." In terms of the waterwheel analogy, it is just the same as saying that water always flows downhill.

Another characteristic of heat is that it takes a certain amount of heat to raise the temperature of a given object a certain number of degrees. For example, a large object is much harder to "warm up" than a smaller one; that is, it takes more heat to do so. In terms of the caloric theory, the temperature corresponds to the "height" of the caloric fluid within the object. The amount of heat needed to raise the temperature of an object one degree Celsius is called the heat capacity of the object. It is not size alone that determines the heat capacity of an object. Objects made of different materials but with the same mass may require different amounts of heat to warm up. This characteristic of a material is given the name *specific heat* and its numerical value is determined by the amount of heat necessary to raise one gram of the material by one degree Celsius.

Water has been chosen to be the standard to which other materials are compared, and its specific heat is defined to be unity. The amount of heat required to raise the temperature of a gram of water one degree Celsius is called one calorie (one "unit" of caloric fluid). Eighteenth-century physicists believed that materials with different specific heats possessed different abilities to hold caloric fluid. The heat capacity of an object is calculated by multiplying its mass by its specific heat.

Another characteristic of materials explained within the framework of the caloric theory of heat is called *latent heat*. Water provides a good example of this concept. As heat is added to water, its temperature rises relatively quickly until it reaches its boiling point. If a thermometer is placed in the water, it will be seen that the temperature quickly rises to the boiling temperature (212°F; 100°C) and then remains there as the evaporation (boiling) begins. Heat is continually added at the boiling temperature while the water vaporizes or turns into steam. Because there is no temperature rise, it may be asked: Where is the heat being added to the water going? The supporters of the caloric theory answered that it was going into a hidden, or latent, form in the water. They did not suggest that it was disappearing.

There is no question that some heat is being added during this time. It is simply going into an invisible form and can be completely recovered by cooling the water. This hidden heat is called the latent heat of vaporization. A parallel phenomenon is observed when heating ice to melt it. Additional heat is required after the ice first reaches its melting temperature to actually melt the ice. This hidden heat is called the latent heat of melting. These latent heats have been observed for all materials as they change from solid to liquid or liquid to gas and are different for different materials. These latent heats are determined as the amount of heat required to melt or vaporize one gram of the material.

It was regarded as significant that exactly all of the heat put into latent heat is released when the material is later cooled. For example, water in the form of steam is more scalding than the same amount of hot liquid water at the same temperature. The latent heat of the steam will be released as it condenses on skin. Something appears to be conserved—that is, not destroyed. This conserved quantity was believed to be the caloric fluid.

There were other phenomena, however, that were not explained sufficiently by the caloric theory; some of these "problem phenomena" led a few physicists to perform investigations which ultimately demonstrated that there is no caloric fluid at all. Perhaps the best-known of these problem phenomena was heat generated by friction. We can warm our hands merely by rubbing them together. There does not appear to be any heat source for the warmth one feels. There is no fire or even a warm object to provide a flow of heat into our cold hands. Where does the heat come from? If somehow heat is being generated where there was none before, we must be (according to the caloric theory) producing caloric fluid. But caloric fluid was believed to be something that cannot be created or destroyed; it can only flow from one object to another. Where, then, does the caloric fluid come from when friction occurs?

The calorists' answer was that a latent heat is released when heat is produced by friction. The claim was that a change of state is involved, just as in changing a liquid to gas, when small particles are ground off an object by friction—during machining, for example. Some calorists said that the material was "damaged" and would "bleed" heat, answers that were never entirely satisfactory. Any small pieces ground off by friction appear to be just tiny amounts of the original material. Eventually it was shown that such grindings have the same characteristics, including the same specific heat, as the original materials, and they certainly had not undergone a change of state.

One of the first people to be impressed by this difficulty with the caloric theory was Count Rumford.* In a famous experiment performed in 1798, Rumford measured the amount of heat developed from friction during the boring of cannon at the royal Bavarian arsenal in Munich. He was greatly impressed with the large amount of heat produced and the small amount of metal shavings that resulted. He was convinced that the tremendous amount of heat being generated could not be from some latent heat. He measured the specific heat of the shavings and found it to be the same as the original metal. Rumford proceeded to demonstrate that, in

---

*Originally Benjamin Thompson, he was a former American colonist and Tory who fled to England during the American Revolution. Ironically, Rumford, who helped overthrow the caloric theory, married the widow of Lavoisier, who had first proposed it.

fact, one could produce any amount of heat one desired without ever reducing the total amount of metal. The metal was simply being slowly reduced to shavings with no change of state involved. He even proceeded to show that heat could be generated without producing any shavings at all. Rumford seriously questioned whether the calorists' interpretation could be correct and suggested that the heat being produced was new heat.

Actually, the calorists readily accepted the results of Rumford's experiments and felt they provided important clues regarding the nature of the caloric fluid. It was said that these experiments showed that caloric fluid must be composed of small, nearly massless particles whose total number in an object is vastly greater than the number ever released by friction. This interpretation is very close to our modern understanding of electricity in terms of atomic electrons. In many ways, Rumford's experiments were simply taken to further the understanding of the caloric fluid.

Besides friction, a second class of phenomena that caused difficulties for the caloric theory was the expansion and contraction of gases. In a little-known experiment in 1807, the French physicist Joseph Louis Gay-Lussac measured the temperature of a gas allowed to expand into an evacuated chamber (one with all the air removed). He measured the temperature of the gas both in its original chamber and in the chamber that had been originally evacuated. The temperatures were found to be the same and equal to the original temperature of the gas. The caloric theory predicted that half the original caloric fluid should be in each chamber (if the two chambers were of equal size). Because temperature was believed to be determined by the concentration of caloric fluid, this should have resulted in the temperature being considerably lower than it was originally. This experiment and others similar to it simply were not understood by the calorists and, in fact, were not understood by anyone for more than thirty years.

The final blow to the caloric theory was provided by the careful experiments of British physicist James Joule in the 1840s. It was well known that heat could do work, and this was understood by the calorists with their waterwheel analogy. Heat did work as it flowed down-temperature just as water does when it flows downhill. But can work create heat? This was one of the most important questions to arise in the entire development of our present-day understanding of heat and energy. The caloric-theory answer was a very definite no. This would be creating new caloric fluid. It would be like creating water by mechanical means, and could not be done—except that is just what Joule's experiments demonstrated.

Joule, encouraged by the work of Count Rumford and others, began to study whether mechanical work could produce heat. His experimental apparatus consisted of a paddlewheel inside a cylinder of water (Figure 4-2). The vanes placed inside the cylinder just barely allowed the paddles to pass and kept the water from swirling with the paddles. Joule wanted to see if one could raise the temperature of the water simply by rotating the paddle. The vanes ensured that the rotating paddle

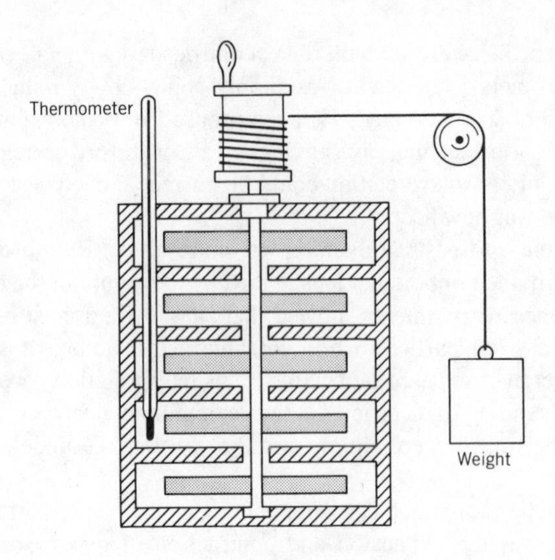

Figure 4-2. Joule's churn. Apparatus used to determine the mechanical equivalent of heat.

could only "agitate" the water, and the microscopic constituents (that is, molecules) would be set into faster motions. Too, if the temperature of the water rose, it must have been because of the mechanical work expended in turning the paddlewheel. Thus if Joule measured a temperature rise in the water, he would have direct evidence that work can produce heat. Finally, one certainly cannot talk about damaging or bruising water to make it "bleed" heat.

Although the temperature rise was rather small, and Joule had to construct his own very sensitive thermometer, he observed a definite increase in the temperature of the water just by rotating the paddle. Joule even set up the experiment to measure just how much heat is produced by a certain amount of mechanical work. Instead of just turning the paddle by hand, he set it up so that a metal cylinder suspended by a rope over a pulley could be allowed to turn the paddle as it fell a measured distance. Earlier in this chapter, work was defined as the product of force multiplied by distance. For Joule's experiment, the force is just the weight of the metal cylinder and the distance is how far it is allowed to fall, so he could measure the mechanical work exactly. He could determine the amount of water in his experiment and, from the temperature rise and the specific heat of water, deduce the amount of heat that must have been generated. Thus Joule determined the mechanical work equivalent of heat. He did this for a number of different materials, such as various types of oils, and he always got the same value for the mechanical equivalent of heat. Joule eventually determined a value which varies less than 1 percent from the modern accepted value. These experiments represent a landmark in the history of science.

There are two extremely important consequences of Joule's experiments. The first is that mechanical work can cause a temperature rise. Work can produce heat. Heat is not conserved. The caloric theory cannot be correct and in fact there is no caloric fluid that can be created or destroyed. We need another theory of heat. Heat must be a form of energy. Recall that mechanical work can produce kinetic energy or potential energy. Joule demonstrated that work can also produce heat. Heat therefore must also be a form of energy.

The second important consequence of Joule's experiments is the clue about how to construct a new theory of heat. Because of the vanes in the cylinder of water, the rotation of the paddles can do nothing to the water except agitate it. This restriction was an aim intended by Joule. Thus it must be concluded that agitated water is warmer water. But what is being agitated in the water? The answer is the molecules of which the water is comprised. These molecules have the familiar chemical symbol H_2O, meaning they consist of two atoms of hydrogen and one of oxygen. It must be molecules that are being agitated, and apparently the more agitated they are the warmer the substance. Temperature must be related to the energy of the microscopic motions of these molecules. Thus heat can only be understood in terms of molecular motions, as described by the molecular model of matter.

As early as 400 B.C. the Greek philospher Democritus suggested that the world was made up of a few basic building blocks, too small for the eye to distinguish, called *atoms*. It was originally believed that there existed only a few different kinds of atoms. Later it was erroneously thought that there was a different kind of atom for evey different material, such as wood, rock, air, and so on. We now know that there are hundreds of thousands of different chemical substances, now called compounds. The smallest amount of a compound is called a *molecule*, not an atom. Molecules, however, are made from atoms, of which there are only about a hundred chemically different kinds. Each atom is associated with one of the elements, from hydrogen to uranium. Molecules and their constituent atoms obey the laws of mechanics and also the laws of conservation of energy and momentum. The dominant force is electrical and not gravitational.

It should be noted that we now know that even the objects identified as atoms are not the fundamental building blocks of nature as originally supposed. Atoms are themselves built up from protons, neutrons, and electrons. It now even appears that protons and neutrons are built up of even smaller units called quarks. In order to understand heat, however, one need not be concerned with the structure of a molecule.

Consider what happens to the molecule of a substance such as water as the substance goes from a gas (steam) to a liquid (water) and finally to a solid (ice). When the substance is a gas, the molecules are essentially not bound to each other at all. Each is free to move in any direction whatsoever and to spin and roll. For a gas at normal pressures (such as the pressure of the atmosphere around us), molecules will not proceed very far in any one direction before colliding with other molecules. They will bounce off each other, obeying Newton's laws of motion and the laws of conservation of momentum and kinetic energy. If we add one gas to another, as by spraying perfume in one corner of the room, the molecules of the new gas will, after many collisions with the molecules of the original gas, diffuse out in all directions. Soon the perfume can be detected across the room as the perfume molecules become completely mixed with the original air molecules.

As the gas is cooled, the molecules move more slowly and usually do not collide with each other with as much speed as when the gas was warmer. When the gas is cooled enough, we know it will condense into liquid. In the liquid state, the molecules are still fairly free to move about, colliding with each other as they move. However, there exists a small force of attraction between all the molecules (which is electrical in origin), and this small force from all nearby molecules adds up, preventing an individual molecule from escaping all the others and leaving the liquid. Thus liquids will remain in a container, even without a lid, for a long time. We do know that most common liquids will slowly evaporate. This is explained by noting that occasionally a molecule will be struck by two or more other molecules in rapid succession and driven toward the surface of the liquid. This gives the molecule an unusually large velocity, which allows it to overcome the net force acting on it from all other molecules, and it will (if there are no intervening collisions) escape from the liquid.

If the liquid is cooled enough, it will finally "freeze" into a solid form. In a solid, the molecules are moving so slowly that the small forces on each one from the nearby molecules hold it in one place. The molecule is no longer free to move around among all the other molecules. The only motion it can have is a vibrational motion about its particular place in the solid.

The molecular description of the states of matter has been verified repeatedly by many careful experiments. No "problem phenomena" are known to exist that question the accuracy of the account of how a material passes from the gaseous to the solid state as the temperature is lowered. (Many of the details of these transitions are not yet fully understood, however.) Thus gases, liquids, and solids are all composed of particles (the molecules) that are moving and are subject to the laws of mechanics.

Many of the earlier discussions can now be understood in detail. Heat input causes the molecules to move faster; heat removal causes them to move more slowly. If the molecules are moving very rapidly, the material is "hot." In fact, the temperature of a gas (whose molecules interact only by elastic collisions) is simply a measure of the average kinetic energy of random translational motion for all the molecules. The average speed of a hydrogen molecule in gaseous hydrogen at normal room temperature has been found to be approximately 4000 mph! (Oxygen molecules, which are sixteen times more massive than hydrogen, at the same temperature and therefore with the same average energy have an average speed of 1000 mph—recall that kinetic energy is calculated from $1/2mv^2$.) Because there about a million billion molecules in one cubic foot of air, gas molecules collide with each other frequently (each molecule is struck by other molecules about 100 billion times per second).

It is interesting to note that one of the first direct verifications of the molecular nature of matter was provided by Albert Einstein, who explained the observed motion of very small solid particles suspended in a liquid solution. This motion, called Brownian movement, is very irregular with the particle moving a short distance in one direction and then quickly in another direction and quickly again in another, and so forth. Einstein explained (and showed quantitatively) that this motion was caused by bombardment of the small particle from the even smaller individual molecules of the liquid.* The same kind of motion can be seen for the dust particles that make a "sunbeam" visible.

One can now also understand why a gas can exert a pressure such as it does in an automobile tire. The individual molecules of the gas are rapidly hitting the sides of the tire and holding it out. The individual molecules are very light, but they are moving very fast and a large number of them hit the wall of the tire every second. The pressure represents the average force exerted by molecules on the wall as they hit and bounce off. Of course, the air molecules on the outside of the tire also exert a pressure on the other side of the walls of the tire at the same time the molecules inside exert their pressure. There are more molecules per unit volume on the inside, however, and thus the inside pressure is greater than the outside pressure by the amount necessary to hold up the vehicle. This extra pressure inside is accomplished by forcing (from an air pump) more molecules per unit volume inside the tire than there are in the air outside the tire. If the temperature of the gas (air) inside the tire were to be raised, the pressure would also increase, because the molecules would have more kinetic energy. Such an increase in pressure occurs when a tire becomes warmer after a long drive on a hot day. The flexing of the tire results in more agitation of the molecules, just as in Joule's experiment.

Specific heat can also now be understood in detail. Recall that the specific heat of a material is the amount of heat required to raise the temperature of one gram one degree Celsius. This means increasing the average kinetic energy of the molecules. Depending on the actual geometric arrangements of the atoms in a molecule or solid, different modes of motion are possible (such modes may include internal vibrations). Thus the specific heats of various compounds will be different because the averaging over the different modes of vibration will be different. Similarly, one can understand latent heats. For example, in order to make a liquid boil, enough heat must be supplied to overcome the net force on an individual molecule produced by all the other molecules still in the liquid state, so

*The particles observed in Brownian motion can be seen with a low-powered microscope. Molecules, however, are so small that they cannot be seen in any significant detail even with the most powerful microscopes presently available. Until Einstein's work many scientists felt that there were no convincing proofs of the existence of molecules. Einstein published this work in the same year (1905) in which he published his first paper on relativity theory (Chapter 6).

we do not see a rise in temperature of the molecules remaining in the liquid. Once enough heat has been absorbed to overcome the bonds between the molecules and the liquid boils into a gas, any additional heat will serve to increase the kinetic energy of the molecules and a rise in temperature will occur.

The kinetic-molecular model of matter reduces the problem of understanding heat and the states of matter to understanding systems of particles with certain forces between them and which obey the laws of mechanics. The success of this model made it appear that Newton's laws of mechanics permit a unified description of the behavoir of all objects, large and small, and in complete detail. Only the development of quantum and relativity theories in the twentieth century (discussed in Chapters 6 and 7) placed limitations on the universality of Newton's laws.

Conservation of Energy

Enough has now been learned to recognize that heat surely is another form of energy, together with kinetic and potential energies. In the discussion of conservative systems (such as a pendulum) it was concluded that the total mechanical energy of the system was almost maintained at a constant value. The fact that the pendulum does eventually come to a stop indicates that the total energy calculated as the sum of the kinetic plus potential energies is not exactly conserved. It seemed likely that the friction at the pivot and between the pendulum bob and the air made it slowly stop.

We have now seen that friction generates heat (Count Rumford's experiment) and that heat is a form of energy. It seems only reasonable to see whether or not the total amount of heat generated by the pendulum is exactly equal to the mechanical energy it loses, but this is a very difficult experiment to perform accurately. One would need to have a way to measure the amount of heat produced and know that it represented the amount of energy that the pendulum had lost. Joule measured how much heat was produced by a certain definite amount of mechanical work and, of course, work produces energy. But how can it be established that some of the energy is not lost? If the heat produced in Joule's experiment could be all converted back into mechanical energy (kinetic or potential), then energy conservation would be proved. But it turns out that heat energy can never be totally converted into another form of energy (this will be discussed in detail in the next chapter). So how did a few scientists originally come to believe that by including heat as another form of energy the total energy of an isolated system is conserved?

Perhaps the first person to become thoroughly convinced that energy is conserved was a German physician, Julius Robert Mayer (1812–1878). Mayer served as a ship's surgeon on a voyage to Java in 1840. He noticed, while treating the members of the crew, that venal blood is redder in the Tropics than in Germany. Mayer had heard of a theory by Lavoisier that body heat came from oxidation in body tissue using the oxygen in blood. Mayer reasoned that because the body needs to produce less heat in the Tropics than in Germany, the blood was redder because less of its oxygen was being removed.

Mayer was intrigued by this phenomenon and continued to think about its implications. He noted that body heat must warm the surroundings. This heat is, of course, from the oxidation of the blood going on inside the body. But there is another way a body can warm the surroundings—by doing work involving friction. Mayer wondered what the source of this kind of heat was. He concluded that it must also be from the oxidation of body tissue by blood, originally. Thus the oxidation process introduces heat into the surroundings both directly by radiation of body heat and indirectly through the link of mechanical work (by the action of muscles). Because the heat in both cases must be proportional to the oxygen consumed, so must be the amount of mechanical work that temporarily takes the place of one part of it. Therefore Mayer believed that heat and work must be equivalent; that is, so many units of one must exactly equal so many different units of the other. Mayer reasoned further that energy originally stored in the food was converted in the oxidation process into energy in the forms of heat and mechanical work. This was the beginning of Mayer's idea of conservation of energy.

All of this was relatively easy for Mayer. It was probably easy because he ignored the caloric theory. He did not consider that heat itself was conserved—that it could not be created where there was none before as demanded by the caloric theory. He assumed that all this heat from the body was new heat. This ignorance of an accepted theory was certainly an asset at first, but it later caused him great difficulty in making his ideas well known. After considerable investigation and thinking, Mayer published a private pamphlet, "Organic Motion in its Connection with Nutrition," in which he recognized various forms of energy. The forms included mechanical, chemical, thermal, electromagnetic, heat, and food. He even proposed, eighty years ahead of his time, that the Sun was the ultimate source of all these energies.

Mayer's proposal that energy is conserved was ridiculed or simply ignored because of his archaic language and use of metaphysics, or sometimes even just incorrect physics. Although he spent years trying to convince scientists of his conclusions, he saw others get credit for his ideas and temporarily lost his sanity from his extreme frustration.

Although Julius Mayer may have been the first scientist (for that is what he became) to recognize the law of conservation of energy, it was the thorough experimental work of Joule, whose paddlewheel experiment was described earlier, that led to the scientific community's acceptance of this idea. William Thomson, Lord Kelvin, a well-known British physicist of the mid-nineteenth century, recognized the importance of Joule's work and gained its acceptance by the Royal Society in London. Kelvin was originally a firm believer in the caloric theory and was aware of Carnot's extensive work on heat engines. Eventually, however, he realized that Joule's work provided a vital clue to the correct analysis of heat engines. The amount of heat transferred to the low-temperature reservoir (the surrounding atmosphere) is less than the amount of heat removed from the high-temperature reservoir (the steam in the condenser chamber) and the difference is found in the mechanical work done. Heat and mechanical work must be two different forms of the same thing: energy. Kelvin's recognition of this fact for heat engines was a very important step in the development of the energy concept.

Joule continued to perform careful experiments and was able to show that electrical energy could be converted to heat with a definite relationship and then eventually that electrical energy and heat could be converted from one to the other. By this time it was already known that certain chemical reactions could generate electrical current and the stage was finally set for recognizing that energy has several forms and can be converted from one of these forms to another.

By the time Joule had finished many of these now-famous experiments he apparently became convinced that energy was something that could not be destroyed but could only be converted from one form to another. For the first few years (starting about 1845) that Joule expounded a theory of the conservation of energy, almost no one would believe he was right. It was not until Kelvin became convinced that Joule's experiments were correct, saw the connection with the new interpretation of heat flow in a steam engine, and then supported Joule's idea that the world began to accept the idea of conservation of energy.

There was also a third "codiscoverer" of the idea of conservation of energy, the German physician Hermann von Helmholtz. In 1847, at twenty-six, he presented a paper entitled "On the Conservation of Force." Although unaware of Mayer's work and hearing of Joule's experiments only near the end of his own work, Helmholtz concluded that energy must have many different forms and be conserved in its total amount. He arrived at this conclusion because he believed that heat, light, electricity, and so on are all forms of motion and therefore forms of mechanical (kinetic) energy. It must have seemed obvious to Helmholtz that kinetic energy was always conserved, not just in elastic collisions but in all collisions, because it went into other forms that were really also kinetic energy. Much of what Helmholtz so boldly suggested was eventually found to be correct, although not all forms of energy can be reduced to kinetic energy. The immediate response to his paper was discouraging, and most historians of science agree that it was the experiments of Joule, backed by the reputation and influence of Kelvin, that eventually convinced the physicists of the mid-nineteenth century that there was indeed a law of conservation of energy.

Let us review what has been discussed thus far. It has been said that heat is a form of energy and that when an account is kept of energy in all its forms, the total amount in an isolated system is conserved. This is the law of conservation of energy. What, then, are all the possible forms of energy?

Basically, for the present discussion, energy can be considered to be in one or more of five major forms: (1) kinetic, (2) heat, (3) potential, (4) radiant, and (5) mass.

Kinetic energy is one-half of Huygens' *vis viva* (or $1/2mv^2$), the energy associated with a body due to its organized motion. Heat has been discussed in some detail, and is energy added to (or taken from) the microscopic modes of (disorganized) motion of the individual molecules of matter. The amount of heat energy is determined by the change in temperature of the object (the average kinetic energy of random microscopic motions of the molecules), and the amount of matter in the object (the number of molecules). Potential energy is energy associated with the shape, configuration, or position of an object in a force field. It was first introduced as gravitational potential energy, but because of the work of Joule and others is now known to have several forms. Some of these forms include (1) gravitational, (2) electrical, (3) chemical (electrical on a microscopic scale), and (4) nuclear.

Whenever energy appears in one of these forms of potential energy, it is possible (although not always practical) to convert the energy completely into kinetic energy or heat energy (which is microscopic random kinetic energy). The fourth major form of energy listed is radiant energy and is electromagnetic radiation such as light, a form that has not been discussed here. For a long time, radiant energy was believed to be a form of heat, but it is not. Radiant energy is also a form of energy that is completely convertible to one of the other forms.

That mass is a form of energy was suggested by Einstein in 1905 (see Chapter 6) and verified some thirty years later when certain nuclear reactions were carefully observed experimentally. Until Einstein's work it was believed that there were separate conservation laws for energy and mass. Newton had included the law of conservation of mass in his *Principia*. In fact, mass and all the other forms of energy are separately conserved to a very good approximation, unless a process including a nuclear reaction or a very high speed is involved. Ordinary chemical reactions such as oxidation involve only an extremely small transformation of mass into energy or vice versa. Because nuclear reactions do not occur naturally on the surface of the Earth (except in radioactive materials which were not noticed until about 1896), the early workers who first established the law of conservation of energy had no reason to recognize mass as another form of energy.

One of the quantities that needs to be determined to characterize a system completely is now seen to be the total energy of the system. One should specify not only the size, mass, temperature, and so on of a system but its total amount of energy as well. This clearly is more appropriate than just indicating the heat content, which was often done before the energy concept was fully developed.

The workings of a system can be analyzed to keep an account of the total amount of energy and how it flows from one form to another. For an isolated system, we now know that if some energy disappears in one form it must reappear in another. This fact can provide an important clue in analyzing a system. Consider a heat engine as an example. If work is done on the engine, energy must have been supplied from somewhere to accomplish it. Conversely, if the engine does work, it supplies a very definite amount of energy to something else. This supplied energy could be kinetic (increasing the speed of a car) or potential (lifting a weight). In the next chapter we will discuss how Rudolf Clausius (1822–1888) was able to reanalyze the workings of a heat engine with an awareness of this law of conservation of energy. He demonstrated that although heat definitely is a form of energy, it is a very special form of energy. As applied to systems, the law of conservation of energy is called the first law of thermodynamics.

Perhaps we should consider for just a moment what this energy, whose total amount in all forms is conserved in an isolated system, actually is. The first thing to be noted is that it is strictly an abstract quantity. Energy is not a kind of material. Each kind of energy is defined by some mathematical formula. Moreover, energy has only a relative nature. Gravitational potential energies are normally measured with respect to the surface of the Earth. Sometimes, however, one might measure how much potential energy an object has with respect to the floor of a room, to the top of a table, and so on. One is only interested in how much potential energy the object has relative to other objects around it. Kinetic energy also has a relative nature, because kinetic energy is measured as the quantity $1/2mv^2$ where v, the velocity, is measured with respect to an arbitrary reference. The chosen reference might be the Earth, the floor of a room, a moving car, the Moon, or whatever seems appropriate for the situation. All potential and kinetic energies are relative.

The next thing to be understood about energy is that only energy changes are significant. There is no need to keep track of all the potential or kinetic energies of an object unless those energies are going to change. It is the *change* in potential energy that shows up as a change in kinetic energy (or vice versa), for example. These changes in energy are especially important when one considers interacting systems. In fact, it is by energy changes that interactions proceed. It can be important to know that if one system loses a certain amount of energy, the other (for only two systems involved) must have gained exactly that amount. Scientists and engineers use the law of conservation of energy continually in analyzing or designing systems involving energy transfers.

The relationship between energy and the physicists' concept of work cannot be overemphasized. Work is defined to be a force acting over a distance. Work is in fact how things are accomplished in the world. Except that thinking is also considered (by most people) to be "work," the physicists' work is involved in all the forms of work of which we are commonly aware. Work produces energy (either kinetic or potential, for example) and energy can do work. From a utilitarian viewpoint perhaps this is our best understanding of what energy is. Energy has the ability to do work; it can cause a force to be exerted over a distance.

Heat was originally thought to be a fluid, but now it is recognized as one of the several different forms of energy. Energy is not (as heat was originally thought to be) any actual substance. Energy is strictly an abstract idea. It is *associated* with motion, or position, or vibrations, and is determined from mathematical expressions for each of its different forms. Energy can be converted into forces acting on objects (Helmholtz referred to the law of conservation of energy as the law of conservation of force before a careful distinction was made between force and energy). It is because energy is conserved that it is such an important concept.

That nature "chooses" to maintain the amount of this strictly abstract quantity at a constant value in the many possible transformations of energy from one form to another is quite amazing and, from a practical viewpoint, important to know.

To conclude this chapter, consider an analysis of a common energy system—the heating of a home by an electrical resistive system—as a demonstration of the concepts of energy. In most parts of the United States, electrical power is produced by the burning of fossil fuels. The energy, therefore, was originally stored as chemical potential energy in the fossil fuel. This energy is converted into heat energy by the burning of the fuel. The heat energy is converted into the kinetic energy of a directed stream of hot gas that exerts pressure on the vanes of a turbine engine, causing it to do mechanical work. The turbine engine in turn drives the shaft of an electrical generator, which converts the work into electrical energy. This electrical energy is then transmitted to the house to be heated. In the house, the electrical energy is converted into both heat energy and radiant energy in a resistive heater unit. This emitted energy is then finally absorbed by the air molecules in a room, giving them more random motion: heat. (Strictly speaking, of course, heat is best defined as energy transferred between two objects as a result of a temperature difference between the two objects. However, the temperature of an object can be related to the random motion of its constituents, at least for the purposes of this book. This is an example of the need, for the purposes of explanation of a complex subject, to oversimplify matters somewhat.)

The law of conservation of energy does not require that the amount of energy finally delivered to the home as heat be exactly equal in amount to the energy originally stored in the fossil fuel. This whole process involves "loss" of energy at every step. The amount finally delivered may be considerably less than originally available. The law of conservation of energy *does* require that all of the energy will go somewhere. Most often some of its escapes from the network as heat and raises the temperature of the surroundings at that point. The law of conservation of energy also requires that more energy never be released from the system at any point than is inserted at that point. Many schemes have been suggested that would try to do just this; that is, release more energy than is put in. Any device that would violate this first law of thermodynamics is known as a perpetual motion machine (of the first kind) because some of the energy released could be used to drive the machine itself. No such scheme has ever been successfully executed, and the first law of thermodynamics stands as one of the most important principles known describing the way nature behaves.

Rudolph Clausius
(American Institute of Physics Niels Bohr Library)

5 Entropy and Probability

Entropy tells it where to go

The idea of energy is so ingrained in everyday life that there is even a Florida theme park, Epcot, devoted to it. However, the energy concept cannot be completely understood until one understands the concept of entropy. Yet entropy is hardly mentioned at Epcot, perhaps because it seems very esoteric. Nevertheless, it is a very important concept, particularly applicable to concerns about energy shortages in modern industrial societies. While it was once thought that boundless resources of energy are available for human use, some people now fear that we may be running out of available energy and that there is a great need for energy conservation. (This is not the same energy conservation expressed in the first law of thermodynamics.) Energy shortages are really shortages of available, useful and "clean" energy, where the meaning of useful and clean are at least partially determined by possible economic, political, and environmental consequences of the conversion of energy to a "useful" form.

The availability of the energy of a system is related to its entropy. Entropy, however, is much more significant than is indicated by its relationship to energy transformations. It is particularly significant for chemistry and chemical processes, and is an important part of the studies of scientists and engineers. For example, the entropy of a substance determines whether it will exist as a solid, liquid or gas and how difficult it is to change from one such state to another. The microscopic interpretation of the entropy concept leads to ideas of order-disorder, organization and disorganization, irreversibility, and probabilities within the kinetic-molecular model of matter. These ideas have been extended and broadened to become significant parts of information theory and communication theory and have been applied to living systems as well. They have also been applied by analogy to political and economic systems. This chapter, however, is mostly about the development of the entropy concept and its applications to energy transformations, reversibility and irreversibility, and the overall direction of processes in physical systems.

One of the major results of the entropy concept is the recognition of heat as a "degraded" form of energy by contrast with other forms of energy. When energy is in the form of heat, it is degraded in the sense that it is partially unavailable for use, a fact that must be taken into account in the design of energy transformation systems.

Only in an isolated system is the total amount of internal energy always the same. As noted in Chapter 4, if a system is not isolated, the energy conservation law is formulated in a different, but equivalent, manner. The state of a system can be characterized or described at least partially in terms of its total internal energy content. The state of the system can be further described in terms of such other properties as its temperature, its size or volume, its mass, its internal pressure, its electrical condition, and so on. These properties, all of which can be measured or calculated in some way, are called *parameters* of the system. If the state of a system changes, at least some of the parameters of the system must change. In particular, it is possible that its total energy content may change.

The law of conservation of energy states that the change in internal energy of the system must be equal to the amount of energy added to the system during the change minus any energy removed from the system during the change. In other words, the conservation law states that it is possible to do bookkeeping on the energy content of a system: any increases or decreases in the energy content must be accounted for in terms of energy (in any form) added to or taken away from the system. The law of conservation of energy used in this manner is called the *first law of thermodynamics*.

The concept of energy and its conservation, although making it possible to set up an equation, sets no limits on the quantities entering into the equation. It gives no idea about how much energy can be added or taken away or what rules, if any, should govern the transformation of energy from one form to another. Nor does the energy concept state what fraction of the energy of a system should be in the form of kinetic energy or heat or potential energy or electrical energy, and so on. In fact, as discussed in the preceding chapter, energy is not a tangible substance at all, nor is it possible to say how much energy a system has in an absolute sense because motion and position are relative. Only changes in energy—that is, *energy transformations*—are significant.

There are indeed rules which govern energy transformations. The concept of entropy and related ideas deals with these rules. Historically, the concept developed from the study of "heat" and temperature and the recognition that heat is a special form of energy that is not completely convertible to other forms. Other forms of energy, however, are completely convertible to heat. Moreover, it seems that in all transformations of energy from one form to another, some energy must be transformed into heat. The word *entropy* was coined by the German physicist Rudolf J. E. Clausius (1822–1888) in 1865 (although he developed the concept in 1854) from Greek words meaning "transformation content."

Entropy, like energy, is also a parameter that can be used to describe the state of a system. The property of a system which is described by the value of its entropy is an abstract property. It is not possible to "feel" or "see" the entropy of a system in the same sense that we can "feel" or "see" other parameters such as temperature, pressure, volume, or mass. Entropy is nevertheless a parameter. It determines how the system's internal energy is distributed and how transformable the internal energy is.

Entropy (and temperature also) is ultimately explained in terms of statistical concepts arising out of the kinetic-molecular theory of matter discussed in Chapter 4. In fact, the full impact and power of the energy concept itself is not appreciated without statistical considerations and the resulting connection of energy with temperature and entropy. The idea of entropy was generalized, through this connection with statistics and probability, to become a useful means of describing relative amounts of organization (order versus disorder) and thus is useful for studying various states of matter (gas, liquid, solid, plasma, liquid crystal, and so on).

Heat and Temperature

To understand the relationship between the concepts of energy and entropy we must first make a clear distinction between heat and temperature. One useful way to do this is to use the caloric theory of heat. (Even though the caloric theory is erroneous, the analogy it makes between heat and a fluid helps introduce some otherwise very abstract ideas.) The relationship between heat flow into an object and its temperature is expressed by the heat capacity of the object. The heat capacity of an object or system is defined as the amount of heat it takes to raise the temperature of the object by one degree (see Chapter 4); that is, an object with a large heat capacity needs a large amount of heat to raise its temperature by one degree. For example, a thimble filled with boiling water can have a higher temperature than a bathtub full of lukewarm water, yet it takes much less heat energy to raise the water in the thimble to its high temperature than the bathtub water to its lower temperature because the thimble has a much smaller heat capacity.

Putting aside the caloric theory of heat, temperature is described in terms of the intensity or concentration of internal molecular energy; that is, the internal molecular energy per unit amount of the substance. The distinction between temperature and heat can be explained in terms of the kinetic-molecular theory of matter by describing temperature as related to the *average energy* of microscopic modes of motion *per molecule* and heat as related to changes in the total energy of microscopic modes of motion of all the molecules. The water molecules in the thimble have more energy on the average than the water molecules in the bathtub, but the thimble needed far less heat than the bathtub because there are far fewer molecules in the thimble than in the bathtub.

Although temperature can be "explained" in terms of concentration of molecular energy, the actual usefulness of the temperature concept depends on two other fundamental aspects of heat recognized in common everyday experience: thermal equilibrium and heat flow. For example, how does a mother decide whether or not a crying baby has a fever? If the child has been fed, has no diaper rash, and is wearing a clean diaper, the mother may notice that its face is flushed and its forehead is warm. She then uses a thermometer to take the baby's temperature.

Almost all thermometers depend on the existence of thermal equilibrium. This means that if two objects or systems are allowed to interact with each other, they will eventually come to have the same temperature. Depending upon the circumstances, this may happen very quickly or very slowly, but eventually the temperature of one or both of the objects will change in such a way that they will both have the same final temperature.

Thus, taking the baby's temperature depends upon the interaction between one object, the baby, and a second object, the thermometer. The mother inserts the thermometer into the baby's rectum and waits for several minutes. While the interaction is going on, the state of the thermometer is changing. The length of the mercury column inside the thermometer is a parameter of the thermometer, and it increases as the temperature of the thermometer increases. When the temperature of the thermometer becomes the same as that of the baby, the length of the mercury no longer changes, and therefore the thermometer is in thermal equilibrium both internally and with the baby. Only after equilibrium is reached may the mother remove the thermometer from the baby and read the scale to determine the baby's temperature.

Actually, the thermometer does not measure the temperature directly, but rather some other parameter is measured. In the case of the rectal thermometer, the parameter measured is the length of the mercury in the glass tube. In other types of thermometers, other parameters are measured. In resistance thermometers, the electrical resistance of the thermometer is the parameter that is measured. In thermoelectric thermometers, the thermoelectric voltage is the parameter measured; whereas in a common oven thermometer, the shape of a coil to which a pointer is attached is the parameter that changes. Almost any kind of system can be used as a thermometer but in all cases the parameter that is actually measured must be related to the temperature parameter in a definite mathematical manner. This mathematical relationship is called an *equation of state* of the system. (In this case, the thermometer is the system.) This simply means that if some parameters of a system change, the other parameters must also change, and the changes can be calculated using the equation of state. The dial reading or scale reading of the thermometer represents the result of solving the equation of state for the temperature. The equation of state of the system being used as a thermometer determines the temperature scale of the thermometer. Whenever the mother reads the thermometer she is in effect solving the equation of state of the thermometer.

Every system has its own equation of state. For example, helium gas at moderate pressures and not too low temperatures has the following equation of state relating three of its parameters: $PV = RT$, where P stands for pressure, V for volume, T for temperature (using the thermodynamic or absolute temperature scale, defined below), and R is a proportionality constant. This equation of state is often called the general gas law or the perfect gas law and is a mathematical relationship between the three parameters mentioned. At all times, no matter what state the helium gas is in, the three parameters—temperature, pressure, and volume—must satisfy the equation. (It is also possible to determine the energy content, U, and the entropy, S, of the system once the equation of state is known.) The equation of state of the system can be drawn as a graph of two of the parameters if the other one is held constant. Figure 5-1 shows graphs of pressure versus volume for helium gas at several different temperatures.

The equation of state and resulting graphs for helium are relatively simple because helium is a gas at all but the lowest temperatures. The equation of state and the resulting graphs of pressure versus volume for water are much more complex because water can be solid, liquid, or gas depending upon the temperature and pressure. Some graphs for water are shown in Figure 5-2.

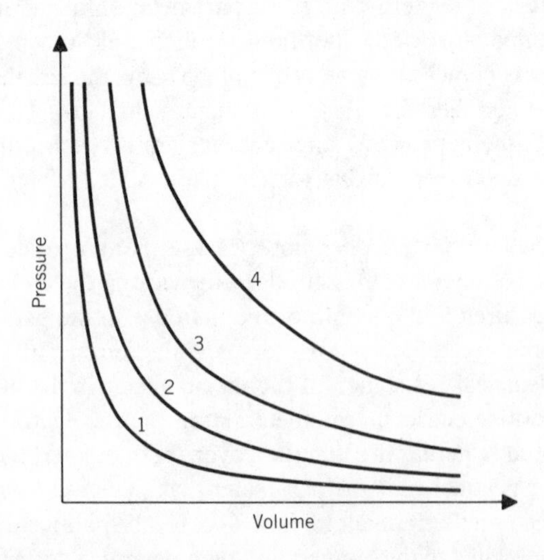

Figure 5-1. Equation of state for helium gas. Each curve is an isotherm (constant temperature curve) showing how pressure and volume can change while the temperature remains constant. The curves 1, 2, 3, 4 are for successively higher temperatures.

Both helium and water could be used as thermometers, and helium is in fact sometimes used as a thermometer for scientific measurements at very low temperatures; except for such special purposes, however, helium is not a very practical thermometric material. Water is useful in establishing certain points on temperature scales, but otherwise is not a practical thermometric substance either. When the barometric pressure is 760 mm (32 inches), water freezes at 0° Celsuis (32° Fahrenheit) and boils at 100° C (212° F). Temperature scales will be discussed further in a later section of this chapter.

It is, in principle, possible to measure the temperature of the system even before the thermometer reaches thermal equilibrium. If two systems at different temperatures are interacting thermally with each other, heat will flow from the higher-temperature (hotter) system to the lower-temperature (colder) system. Thus the mother can tell that her baby has a fever by feeling its forehead. If the baby has a fever, the forehead will feel hot; that is, heat will flow from the baby's forehead to the mother's hand. If, on the other hand, the baby's temperature is well below normal, the forehead will feel cool; that is, heat will flow from the mother's hand to the baby's forehead. This very fact that temperature differences can cause heat to flow is very significant in understanding the concept of entropy and how heat energy can be transformed into other forms of energy. Moreoever, this same fact underlies the correct definition of the thermodynamic (sometimes called absolute) temperature scale.

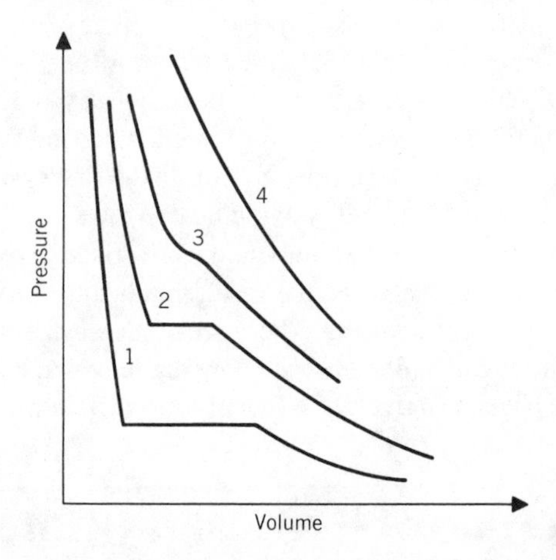

Figure 5-2. Equation of state for water. Each curve is an isotherm showing how pressure and volume can change while the temperature remains constant. The curves 1, 2, 3, 4 are for successively higher temperatures. The flat parts of curves 1 and 2 cover a range of volumes for which part of the water is liquid and part gas. Water is completely liquid to the left of the flat parts and completely a gas to the right. Curve 3 is at a temperature (374° Celsius) for which it is impossible to tell the difference between liquid and gas, whereas at all higher temperatures water can only be a gas.

The Natural Flow of Heat

Everyone knows that heat flows from hotter objects to colder ones. If an ice cube is dropped into a cup of hot tea, the ice cube becomes warmer, melts into water, and the resulting water becomes warmer while the tea becomes cooler. The heat "flows" out of the hot tea into the cold ice cube. As already discussed in Chapter 4, this was easily explained by the caloric theory as being due to heat flowing "downhill." Of course, the caloric theory is wrong, and so it is not possible to explain so simply why heat flows from higher-temperature bodies to lower-temperature bodies and not vice versa. In terms of the distinction between heat and temperature made above, energy flows from high concentrations (high temperature) to low concentrations (low temperature); in other words, energy "spreads out" by itself. This fact of one-way heat flow is the essence of the *second law of thermodynamics*. Like Newton's laws of motion or the law of conservation of energy, this law is accepted as a basic postulate. There are several different ways of stating the second law, all of which can be shown to be equivalent.

In terms of the natural flow of heat, the second law of thermodynamics is stated as follows: *In an isolated system, there is no way to systematically reverse the flow of heat from higher to lower temperatures*. In other words, heat cannot flow "uphill" overall. The word *overall* is very important. It is, of course, possible to force heat "uphill" using a heat pump, just as water can be forced uphill using a water pump. An ordinary refrigerator and air conditioner are examples of heat pumps. However, if a refrigerator is working in one part of an isolated system to force heat "uphill," in some other part of the system more heat will flow "downhill" or further "downhill," so that overall the net effect in the system is of a "downhill" flow of heat. In general, if a system is not isolated, a second system can interact with it in such a way as to make heat flow "uphill" in the first system, but there will be at least a compensating "downhill" flow of heat in the second system.

Although the second law of thermodynamics as stated above seems obvious and even trivial, it leads ultimately to the result that heat is a form of energy not completely convertible to other forms. There are other ways of stating the second law of thermodynamics which demonstrate how fundamental it is for understanding the nature of the physical universe. A few of these different ways are discussed below.

Transformation of Heat Energy into Other Forms of Energy

Long before heat was recognized as a form of energy, it was known that heat and motion were somehow related to each other. The friction that is often associated with motion always results in generation of heat. Steam engines, which require heat, are used to generate motion. By the middle of the nineteenth century it was recognized that a steam engine is an energy converter, a system that can be used to convert or transform heat energy into some form of mechanical energy—work, potential energy, kinetic energy—or electrical energy or chemical energy, and so on.

As discussed in the preceding chapter, even before the true nature of heat as a form of energy was understood, Carnot realized that the steam engine was only one example of a heat engine. Another familiar example of a heat engine is the modern gasoline engine used in automobiles. Carnot realized that the essential, common operating principle of all heat engines was that heat flowed through them. He pointed out that the fundamental result of all of Watt's improvements was more effective heat flow through the engine.

As already noted in Chapter 4, Carnot visualized the heat engine as being like a waterwheel, in that caloric fluid had to flow through it from a high temperature (called the heat source) to a low temperature (called the heat sink). In the spirit of Plato, in an effort to get at the "true reality" Carnot visualized an *ideal heat engine* whose operation would not be hampered by extraneous factors such as friction or unnecessary heat losses. He therefore considered that this ideal engine would be reversible (could move the fluid uphill) and would operate very slowly. (Here he was doing the same thing as Galileo, who in his studies of falling motion used inclined planes and rolling balls to slow the action and get rid of friction and resistance effects.)

The principle of the engine was that heat had to flow *through* the engine to make it operate, and it was necessary to ensure that all the available heat would flow through the engine. This is shown diagramatically in Figure 5-3. In Figure 5-3a all the heat from the heat source flows through the engine. In Figure 5-3b some of the heat from the source is diverted, bypassing the engine, going directly to the heat sink, and thus is "wasted."

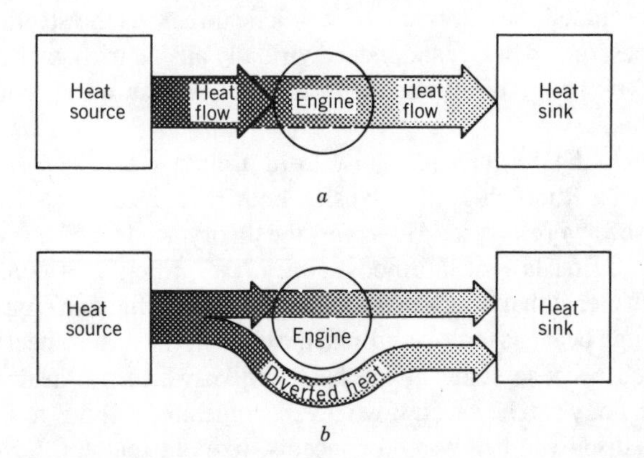

Figure 5-3. Schematic representation of heat flows for a heat engine. (a) All of the heat from the source passes through the engine. (b) Some of the heat from the source bypasses the engine.

Carnot then pursued in detail his analogy between a heat engine and a waterwheel or water mill through which water flows, causing rotation of a shaft and thereby operating attached machinery. In fact, because the caloric theory considered that the heat fluid flowed from high temperatures to low temperatures just as water flows from high levels to low levels, the analogy was very close. The heat engine could be considered to be a heat wheel, which could be designed just like a waterwheel. (Carnot's father had made many suggestions for improving the design of waterwheels, and Carnot was influenced by these.)

Waterwheels operate by virtue of having water fall from a high level to a low level, and similarly heat wheels operate by virtue of having heat flowing from a high temperature to a lower temperature. An efficient waterwheel should be designed to take advantage of the full height of the waterfall and to make full use of all the water coming over the fall. Similarly the heat wheel should take advantage of the full temperature difference between the heat source and the heat sink, and all of the heat should flow through the wheel.

This requires that ideally a heat engine must satisfy several conditions: (1) Heat must be taken in only when the engine is at the temperature of the high temperature reservoir. (2) The engine must be thoroughly insulated while its temperature is changing from high to low, so that no heat will leak out (be diverted) and its motive force be lost. (3) The heat must then be discharged at the temperature of the low temperature reservoir, so that (4) no heat will be carried back "uphill."

Carnot concluded that the efficiency (the relative effectiveness of the engine in providing motion or work) of his ideal heat engine would depend *only* on the difference in temperature of the two reservoirs and not on whether it was a steam engine or any other kind of engine. Carnot did consider that a heat engine could be used which had some other "working substance" than water, as used in the steam engine. He concluded, however, that for practical reasons the steam engine was probably the best one to build, and indeed virtually all electric generating stations which depend on heat (coal-fired, oil-fired, nuclear) use water as the working substance.

About 1850, the English physicist Lord Kelvin (see Chapter 4) and the German physicist Rudolph J. E. Clausius both recognized that Carnot's basic ideas as published were correct. However, the theory needed to be revised to take into account the first law of thermodynamics (the principle of conservation of energy), which meant that heat was not an indestructible fluid but simply a form of energy. Although heat has to flow through the engine, the waterwheel analogy has to be discarded because some of the heat, unlike water, is "destroyed." (The waterwheel analogy can be saved if we invent some other "fluid" to flow through the engine, as discussed below. Old concepts, like old soldiers, never die, they just fade away.) The amount of heat flowing out of the engine into the low-temperature reservoir is less than the amount of heat flowing into the engine from the high-temperature reservoir, the difference being converted into work or mechanical energy. In other words, when a certain amount of heat energy flows into the engine from the high-temperature reservoir, part of it comes out as some other form of energy and the rest of it flows as heat energy into the low-temperature reservoir. All the energy received must be discharged, because after the engine goes through one complete cycle of operation it must be in the same thermodynamic state as it was at the beginning of the cycle. Its internal energy parameter, U, must be the same as it was at the beginning. Therefore, according to the first law of thermodynamics, all the energy added to it (in the form of heat from the high-temperature reservoir) must be given up by it (as work or some other form of energy and as heat discharged to the low-temperature reservoir). This is shown schematically in the energy flow diagram of Figure 5-4.

The essential correctness of Carnot's ideas, according to Clausius, lay in the fact that at least part of the heat flowing into the engine *must* flow on through it and be discharged. This discharged heat cannot be converted by the engine to another form of energy. In other words, it is not possible to convert heat entirely into another form of energy by means of an engine operating in a cycle. To do so would violate the second law of thermodynamics, which states that heat cannot flow "uphill" in an isolated system. Figure 5-5 shows schematically what would happen if an engine could convert heat entirely into mechanical energy.

Suppose such an engine were being "driven" by the heat drawn from a heat reservoir at a temperature of 30° Celsius (86° Fahrenheit). The assumption is that the engine draws heat from the reservoir and converts it completely into mechanical energy. This mechanical energy could then be completely converted into heat once again, as Joule was able to show by his paddlewheel experiment. In particular, the mechanical energy could be used to stir a container full of water at 50° Celsius (122° Fahrenheit), and would be completely converted into heat at 50° Celsius. But that mechanical energy had originally been heat energy in the 30° reservoir. Thus the net result of all this would have been that heat had been transferred from the 30° reservoir to the hotter 50° reservoir with no other change, because the engine working in a cycle returns to its original state. The "driving force" for the entire process was the heat in the 30° reservoir. Heat would have "flowed uphill" by itself—which is impossible according to the second law of thermodynamics.

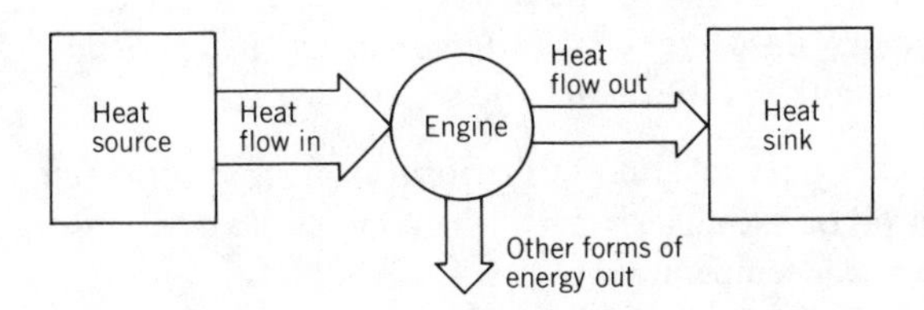

Figure 5-4. Energy flow for a heat engine. Energy input as heat; energy ouput as heat and work or other useful forms of energy.

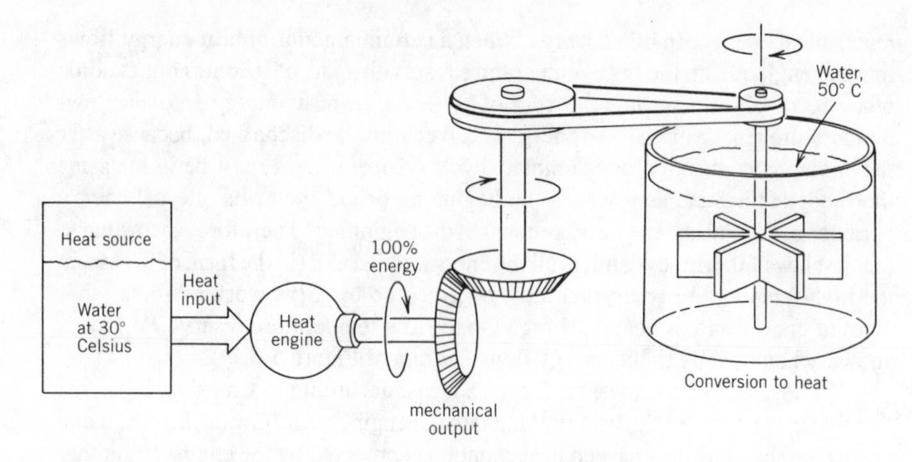

Figure 5-5. Schematic of an impossible machine. Heat taken from a large body of water at 30° Celsius is completely converted into mechanical energy. The mechanical energy is then reconverted into heat by stirring a container full of water at 50° Celsius with the resulting overall effect that heat flowed "uphill."

The flaw in this process is in the assumption of total transformation of heat into mechanical energy. Some heat must move "downhill" untransformed to make it possible for any portion of the heat to be transformed into mechanical energy, which is retransformed into heat energy at a higher temperature than the reservoir from which it was originally derived. Overall there must be a net "downhill" movement of heat. The specific examples of entropy changes to be discussed below will show how this overall downhill movement of heat is calculated.

Sometimes the second law of thermodynamics is stated specifically in terms of these limitations on the convertibility of heat to another form of energy. As just shown, such statements are equivalent to the original statement in terms of irreversibility of heat flow:

1. *It is impossible to have a heat engine working in a cycle that will draw heat from a heat reservoir (source) without having to discharge heat to a lower temperature reservoir (sink).*

2. *It is impossible to have a heat engine working in a cycle that will result in complete (100%) conversion of heat into work.*

Whenever energy is released in the form of heat at some temperature (by burning coal or oil or by nuclear reaction), some of the heat must eventually be discharged to a lower-temperature reservoir.

Any violation of the second law of thermodynamics is called "perpetual motion of the second kind." If it were possible to build a perpetual motion machine of the second kind, some beneficial results could be obtained.

For example, an oceangoing ship could take in water from the surface of the ocean, extract heat from the water, and use that heat to drive the ship's engines. The result would be that the water taken in would be turned into ice and the ship propelled forward. The ice would be cast overboard, the ship would draw in more ocean water, and the process would continue. The ship would not need any coal, fuel oil, or nuclear energy for its operation. The only immediate by-products of the operation of the ship would be the ice, plus various previously dissolved salts and minerals that would have crystallized out of the ocean water during the heat-extraction process. All of these could be returned to the ocean, with the only net result being that the ocean surface is very slightly, essentially insignificantly, colder.

An even more spectacular example can be conceived. Imagine electricity-generating stations located on the banks of the various polluted rivers of the world: the Cuyahoga in Ohio, the Vistula in Poland, the Elbe in Germany and Czechoslovakia, the Ganges in India, and the like. These stations would take in polluted river water and extract heat from the water for conversion into electrical energy, thereby lowering the temperature of the water to below its freezing point. The process of freezing the water results in a partial separation of the pollutants from the water. The now-cleaner frozen water is allowed to melt and is returned to the river, and the separated pollutants can be either reclaimed for industrial or agricultural use or simply buried. The result of all this would be a significant improvement of the environment as a by-product of the generation of electrical energy.

Neither of these schemes will work because, although they would take heat from an existing heat reservoir, they do not discharge any heat to an even lower-temperature heat reservoir to conform to the requirement that heat must flow "downhill." Unfortunately, the second law of thermodynamics complicates the solution to environmental and energy problems.

Efficiency of Heat Engines

The efficiency of any engine is simply the useful energy or work put out by that engine divided by the energy put into the engine to make it operate. In a heat engine, coal or oil or gas might be burned to generate the energy input in the form of heat. This is called Q_H. Suppose that in a particular case 50,000 kilocalories (a common unit of heat) were released by the combustion of fuel to be Q_H. Suppose that of this amount 10,000 kilocalories were converted into electricity. This is called W, for work. (In the following discussion, the word *work* is used for any form of energy other than heat. As already pointed out, work is completely convertible to any other form of energy in the absence of friction.) The fractional efficiency is $W/Q_H = 10{,}000/50{,}000 = 1/5 = 20$ percent. Because only 10,000 kilocalories were converted into electricity, 40,000 kilocalories (the remainder) were discharged into a lower-temperature reservoir. This is called Q_C.

Equivalently, the fractional efficiency can also be calculated from $1 - Q_C/Q_H$. In the example at hand, $1 - Q_C/Q_H = 1 - 40{,}000/50{,}000 = 1 - 0.8 = 0.2 = 20$ percent. This relationship is true for any heat engine, ideal or not. The efficiency of the large Newcomen engines was only about .5 percent and for Watt's improved engine was about 3 or 4 percent. The fractional efficiency of the heat engines in a modern electric generating station is typically 40 percent. Table 5-1 lists the efficiencies of several types of engines. The temperatures given are based on the Kelvin scale, discussed in the next section.

The second law of thermodynamics as given above states that even the very best engine conceivable—the ideal heat engine—cannot be 100 percent efficient. How efficient, then, can this best heat engine be and exactly why is it not fully efficient? These are important practical questions, because the ideal heat engine sets a standard against which other practical engines can be measured. If a particular engine comes close to the ideal efficiency, then its builders will know that they cannot improve its efficiency much more; on the other hand, if the engine

Table 5-1 Efficiencies of Some Heat Engines

Type of Engine	Hot Reservoir Temperature (Kelvin)	Cold Reservoir Temperature (Kelvin)	Efficiency (percent)
Newcomen engine	373	300	1/2
Watt's engine	385	300	3 - 4
	1500	300	80
Ideal Carnot	1000	300	70
	811	311	62
Rankine	811	311	50
Actual steam turbine power plant	811	311	40
Binary vapor cycle	811	311	57
Gasoline engine with Carnot efficiency	1944	289	85
Ideal Otto gasoline engine	1944	289	58
Actual gasoline engine	-	-	30
Actual diesel engine	-	-	40

has only one half or less of the ideal efficiency, perhaps it can be improved significantly. The ideal efficiency is sometimes called the Carnot or *thermodynamic efficiency*.

Even though he used the erroneous caloric theory, Carnot did list the correct steps in the cycle of the ideal heat engine. Therefore, the ideal heat engine is called the Carnot engine and the cycle is called the Carnot cycle for an engine operating between two heat reservoirs. The essential characteristics of this cycle are shown schematically in Figure 5-6, which also shows how the thermodynamic parameters are involved in the engine operation. In the figure the flame and the ice represent the heat source and the heat sink.

Step 1. Isothermal heat intake. (*Isothermal* means constant-temperature.) The engine is at the temperature T_H of the hot reservoir and slowly takes in heat while its temperature is held constant. In order to do this, some other parameter of the engine system must change, and as a result the engine may do some useful work. (In Figure 5-6a the piston is pushed out, increasing the volume of the cylinder.)

Step 2. Adiabatic performance of work by the engine. (*Adiabatic* means no heat can enter or leave.) The engine is completely thermally insulated from the heat reservoirs and the rest of the universe, and does some useful work. The performance of work means that energy (but not heat) leaves the engine, and therefore the internal energy content of the engine is less at the end of this step than at its beginning. Because of the equation of state of the working substance of the engine, the temperature of the engine decreases. This step is terminated when the engine reaches the same temperature T_C as the low-temperature or cold reservoir. (In Figure 5-6b the piston is still moving outward.)

Step 3. Isothermal heat discharge. The engine is now at the temperature T_C of the cold reservoir, the thermal insulation is removed, and heat is slowly discharged to the low-temperature reservoir while the temperature of the engine remains constant. Again, in order to do this, some other parameter of the engine must change as a result of work being done on the engine. It is at this step and the next that some of the work taken out of the engine must be returned to the engine. These two steps, which take over the function of Watt's condenser and are necessary for an engine working in a cycle, can be regarded as the steps where nature reduces the efficiency from 100 percent to the thermodynamic values. (In Figure 5-6c the piston is now driven inward by the flywheel and connecting rod, decreasing the cylinder volume.)

Step 4. Adiabatic temperature increase of engine. The engine is again completely thermally insulated from the heat reservoirs and the rest of the universe. Work is done on the engine, usually by forcing a change in one of its parameters, resulting in an increase of temperature until it returns to the starting temperature, T_H, of the hot reservoir. The internal energy content of the engine is now the same as at the beginning of Step 1 and the engine is ready to repeat the cycle. (In Figure 5-6 the piston continues its inward motion.)

Interestingly enough, there may be portions of an engine cycle in which heat is converted 100 percent to work. Step 1 of the Carnot cycle, shown in Figure 5-6a, is such a portion. However, in order for the engine to continue working it must go through a complete cycle. In the process of going through the complete cycle the second law of thermodynamics is satisfied in Steps 3 and 4 by putting back into the engine a portion of the work output from Steps 1 and 2, some of which is

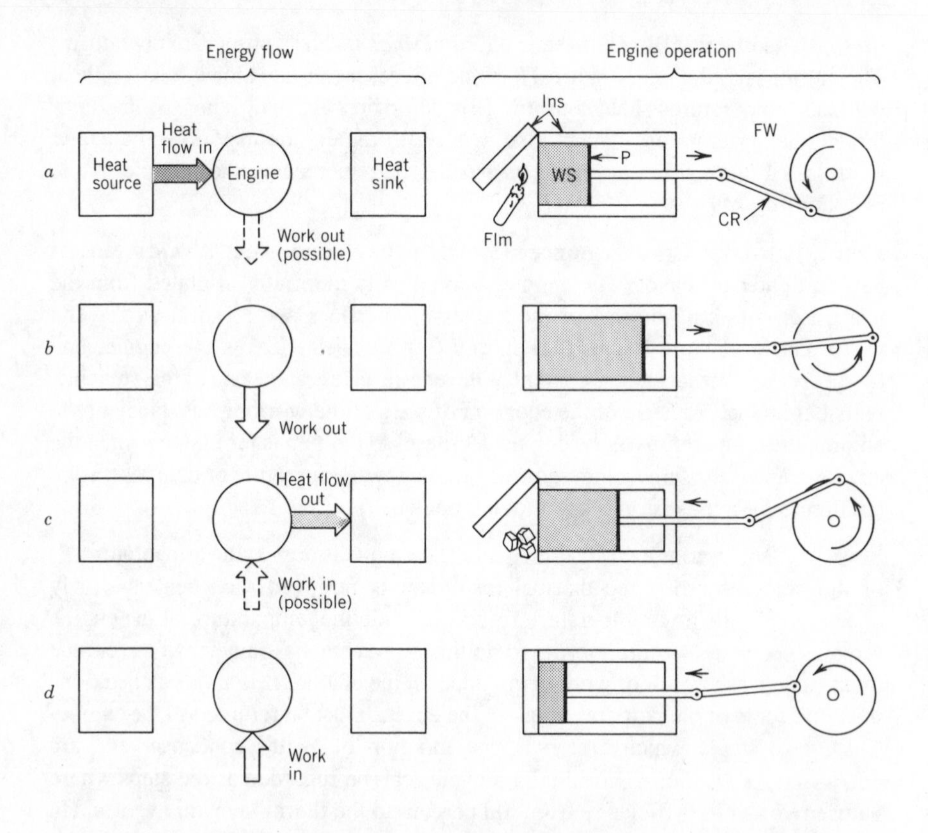

Figure 5-6. The Carnot cycle and associated energy flows. The left side of the figure shows the energy flow associated with diagrams on the right side of the figure. *P*, piston; *WS*, working substance; *CR*, connecting rod; *FW*, flywheel; *Ins*, insulation; *Flm*, flame; *I*, ice. (a) Isothermal heat intake from hot reservoir, piston moving outward, work done by engine. (b) Adiabatic work by engine, piston moving outward. (c) Isothermal heat discharge to cold reservoir, piston moving inward as work is done on the engine by the flywheel. (d) Adiabatic temperature increase, piston moving inward as more work is done on engine by the flywheel.

discharged to the low-temperature reservoir (Figure 5-6c) and the rest used to return the engine to the starting point of the cycle (Figure 5-6d). This is true for all engines, regardless of their construction or operation.

Carnot had originally reasoned that the efficiency of the Carnot cycle should be proportional to the *absolute* temperature difference between the hot and cold reservoirs, $T_H - T_C$. When Clausius reanalyzed the cycle to allow for conservation of energy he showed that, if the proper temperature scale is used (the so-called absolute temperature scale), the efficiency would be equal to the *fractional* (percent) difference in temperature of the two reservoirs relative to the temperature of the hot reservoir; that is, the efficiency equals $(T_H - T_C)/T_H = 1 - T_C/T_H$.

Using Clausius' formula for the Carnot or thermodynamic efficiency, it is possible to calculate the efficiency that could be expected if all friction, irreversibilities, and heat losses were eliminated from a modern electricity-generating plant and it were operated in a Carnot cycle. The results of such calculations, given in Table 5-1, show that actual modern generating stations do get reasonably close to the Carnot efficiency for the temperatures at which they operate.

The only way to increase the theoretical possible efficiencies of heat engines is either by increasing T_H or decreasing T_C. On the planet Earth, T_C cannot be lower than 273 Kelvins (32° Fahrenheit) because that is the freezing temperature of water, and present-day power plants use cooling water from a river or ocean for their low-temperature reservoirs. Theoretically, it would be possible to use some other coolant, say antifreeze (ethylene glycol) as used in automobile engines, which remains liquid at temperatures much lower than the freezing point of water. Then a refrigeration system would be used to lower the temperature of the antifreeze in order that it can be a very low-temperature reservoir. However, energy is required to operate the refrigeration system. It can be shown that the amount of energy that would have to be taken from the power plant to operate the refrigeration system would more than offset the energy gain from the increased efficiency resulting from the lower value of T_C. In fact, the net effect of installing the refrigerator would be to *decrease* the overall efficiency below the Carnot efficiency.

Of course, one can imagine building an electricity-generating station on an orbiting satellite in outer space, where T_C is only a few degrees absolute. It would then be necessary to consider how the electricity generated at the station would be transmitted to Earth.

Alternatively, it is necessary to consider the possibility of raising the temperature T_H of the hot reservoir. This temperature cannot be higher than the melting temperature of the material out of which the heat engine is built. Actually, T_H will have to be considerably less than the melting temperature since most materials are rather weak near their melting points. If T_H were about 1500 Kelvins (the melting point of steel is 1700 Kelvins) and T_C were 300 Kelvins, the Carnot efficiency is calculated to be $1 - 300/1500 = 0.8 = 80$ percent. Because of the need for safety factors (to keep boilers from rupturing under the high pressure associated with high temperatures), T_H would be more reasonably set at 1000 Kelvins, giving a Carnot efficiency of 70 percent. An actual engine might not achieve much more than about three quarters of the Carnot efficiency, yielding a maximum feasible efficiency in the range of 50 to 55 percent. It can thus be expected that about half of the fuel burned in the best practical heat engine operating between two "reasonable" temperature reservoirs will be wasted so far as its convertibility into "useful" work is concerned. (See the data in Table 5-1 for more specific examples.)

An idea has recently been proposed that is in effect a three-reservoir heat engine to obtain more efficient transformation of the heat liberated at high temperatures. Some heat is added to the system at an intermediate temperature between T_H and T_C, for example, during Step 4 outlined above for the Carnot cycle. This heat added at an intermediate temperature then helps bring the engine temperature back up to T_H so that less work needs to be put back into the engine. Step 4 would no longer be adiabatic. The result is that the heat entering the system at T_H is then more effectively transformed into some other form of energy; however, the heat added at the intermediate temperature is transformed less effectively. Overall, the efficiency is less than Carnot efficiency if all sources of heat are taken into account. In effect, Q_C, the heat discharged to the low-temperature reservoir, comes mostly from the intermediate-temperature reservoir rather than from the high-temperature reservoir. The advantage of the idea as proposed is that this intermediate-temperature reservoir would be a source of heat which required that no additional fuel be burned; that is, the source would be either solar or geothermal heat or some low-grade heat source that is already available for reasons other than energy conservation. (See the article by J. R. Powell and colleagues listed in the references for this chapter.)

The ideas involved in heat-engine theory are almost universal. The many varieties of heat engines can be classified according to several categories. For example, there are external-combustion engines (such as reciprocating steam engines, steam turbines, and hot-air engines) and internal-combustion engines (such as the various types of gasoline and diesel engines, jet engines, and rocket engines). Any engine that involves the burning of some fuel or depends on temperature in order to operate, or in any way involves the liberation or extraction of heat and its conversion to some other form of energy, is a heat engine. Even the Earth's atmosphere is a giant heat engine that converts heat from the Sun into other forms of energy. (Wind energy is kinetic energy of systematic motion of air molecules, and thunderclouds contain electrical energy.) All heat engines are subject to the same rules concerning their maximum theoretical efficiency, which is completely governed by the theory developed by Carnot and corrected by Clausius. As already stated, the ideal efficiency (the Carnot efficiency) for all these engines depends only on the temperatures of their two heat reservoirs and is independent of anything else or any details or features of engine construction.

Other than for theoretical purposes, an engine built specifically to conform as closely as possible to the Carnot cycle is not always practical. There is, in fact, a story about a ship's engine that was built to have an efficiency close to that of a Carnot engine. When it was installed in the ship for which it was designed, however, it was so heavy that the ship sank! (See Chapter 18 of Morton Mott-Smith's book, listed in the references.)

The Thermodynamic or Absolute Temperature Scale

Thermometers and their scales were described above as depending upon temperature equilibrium and the equation of state of the thermometer system. It was also indicated that temperature scales could alternatively be defined in terms of the heat flow between two different temperatures. The Carnot cycle makes it possible to do just that. Instead of using the physical parameters of a thermometric substance such as mercury, it is possible to use the properties of heat and heat transformation. In other words, energy can be used as a thermometric "substance."

Imagine two bodies at different temperatures. Imagine further a small Carnot engine that would use these two bodies as its heat reservoirs. According to the Carnot-engine theory, the ratio of the heat flowing out of the engine to the heat flowing into the engine, Q_C/Q_H, should depend only on the temperatures of the two bodies. The British scientist Lord Kelvin proposed that the temperature of the two bodies T_C and T_H should be defined as proportional to their respective heat flows; that is, $T_C/T_H = Q_C/Q_H$. The temperature scale so defined is called a *thermodynamic* or *absolute temperature scale*. For example, suppose that a particular body is known to be at a temperature of 500 absolute and that we wish to know the temperature of some cooler body. It might be imagined that a small Carnot engine is connected between the two bodies and the heat flow in, Q_H, and the heat flow out, Q_C, might be determined for this engine. If the ratio of Q_C/Q_H were 1/2 then, according to Kelvin, the ratio of T_C/T_H should be 1/2 also. Thus the temperature of the cooler body is half that of the hotter body or 1/2 of 500; that is, 250 absolute.

The absolute temperature scale is defined above only in terms of ratios. To actually fix the scale it is necessary also to specify a size for the units of temperature (degrees). If there are to be 100 units between the freezing and boiling points of water (as in the Celsius temperature scale), on the absolute scale the freezing temperature of water is 273 and boiling temperature is 373. This absolute temperature scale is called the Kelvin scale. (In common usage the units on a temperature scale are usually called degrees; by international agreement, however, the units on the Kelvin scale are called Kelvins.) If there are to be 180 degrees between the freezing and boiling points of water (as in the Fahrenheit scale), the freezing temperature of water is 491° and the boiling point is 671°. This absolute scale is called the Rankine scale.

The Third Law of Thermodynamics

What would happen if the low-temperature reservoir were at absolute zero? Since the Carnot efficiency of a heat engine is equal to $1 - T_C/T_H$, the Carnot efficiency would then be 100 percent. Such reservoirs do not normally exist, however. As already pointed out, the energy expended in the attempt to make a cold reservoir at zero absolute temperature would be greater than the extra useful energy obtained by operating at zero absolute temperature. Nevertheless, it would be interesting to see whether the theory would be valid, even at that low temperature. Thus it is necessary to examine the possibilities of making such a reservoir.

It turns out that in order to make a low-temperature reservoir, the effectiveness of the refrigerator used to achieve the low temperature becomes less as the temperature goes down, and it takes longer and longer to reach lower temperatures as the temperature gets closer to absolute zero. In fact, this experience is summarized in the third law of thermodynamics: *It is not possible to reach the absolute zero of temperature in a finite number of steps*. This means that it is possible to get extremely close to absolute zero temperature, but it is always just beyond being attained. In other words, "absolute zero can be achieved, but it will take forever."*

The three laws of thermodynamics are sometimes summarized humorously in gambling terms. Comparing heat energy with the stakes in a gambling house, the laws then become:

1. You can't win; you can only break even.

2. You can only break even if you play long enough and are dealt the exactly correct set of cards.

3. You should live so long as to get the right cards.

*Strictly speaking, it is preferable to say "zero Kelvin" rather than "zero absolute" and "thermodynamic temperature" rather than "absolute temperature," because the temperature scale is based on heat flow, which is a dynamic process. The use of terms such as *absolute zero* can cause confusion, because in certain circumstances the question of negative absolute temperature arises. For further details, see Chapter 5 of the book by Mark Zemansky listed in the references for this chapter.

Energy Degradation, Unavailability, and Entropy

Even though energy is neither created nor destroyed but only transformed, the concept of "lost" or "degraded" energy is useful. If a given heat engine is only 40 percent efficient, then the other 60 percent of the heat generated in operating the engine is not converted to "useful" forms but is discharged to the low-temperature reservoir. Moreover, this heat cannot be recycled through the engine for another attempt to convert it into a useful form. Any heat that is discharged to a low-temperature reservoir is wasted in that it is no longer available for transformation to other forms of energy (unless there is a still-lower-temperature reservoir accessible for use). The energy is said to be degraded. Indeed, a given amount of energy in the form of heat is degraded in comparison with the same amount of energy in some other form, because it is not completely convertible. Thus whenever energy is transformed into heat it becomes degraded.

Moreover, if energy in the form of heat flows out of a high-temperature reservoir directly to a lower-temperature reservoir it will become much more degraded than if it were to flow through a heat engine, where at least part of it is transformed to another form to be available for further transformations. There are two characteristics which can make the energy of a body somewhat unavailable or degraded: (1) that it be in the form which is described as heat and (2) that the lower the temperature of the body the more unavailable its heat energy.

As already mentioned, Clausius sought to describe the availability for transformation of the energy content of a system by coining the word *entropy* from Greek roots meaning transformation content. In fact, he used the word to describe the unavailability of the system's energy for transformation. Thus the *greater the entropy*, the *less available the internal energy* for transformation if a heat engine were set up within the system.

Clausius was greatly impressed by Carnot's ideas. Although he realized that the calorists' concept of heat as an indestructible fluid was wrong, he also realized that an abstract "fluid" (more abstract than either caloric or energy) could be said to flow through an ideal engine and not be destroyed. Entropy is this fluid. Entropy, like energy, is a relative concept, and thus it is changes in entropy that are significant. Like energy, entropy cannot be destroyed; however, unlike energy, it can be created, as will be discussed below. In Carnot's original analysis no heat was "lost" in the ideal heat engine, although it could be lost in any other engine. In Clausius' refined analysis, as will be shown below, no entropy is "created" in the ideal engine, although it can be created in any other engine.

The nature of entropy changes can be conveniently illustrated by numerical examples. In some cases, changes in the entropy of a body or system are calculated in terms of the amount of heat added to or taken away from the body or system per unit of absolute temperature. If a body that is initially at equilibrium at some temperature has some heat added to it, then by definition the entropy of the body increases by an amount equal to Q/T—that is, the amount of heat added Q, divided by the absolute temperature T of the body. (This definition of entropy change, like the definition of kinetic energy as $1/2mv^2$, is presented here without justification.) For example, if a body initially at 300 Kelvins (about room temperature) receives 5000 kilocalories of heat, its entropy change is $+5000/300 = +16.67$ kilocalories per Kelvin. On the other hand, if heat is taken away from a body, its entropy decreases. Thus if a body at 600 Kelvins loses 5000 kilocalories of heat, its entropy change is $-5000/600 = -8.33$ kilocalories per Kelvin. (If the temperature of the body is changing, its entropy change can be calculated from the area underneath a graph of heat versus $1/T$.)

A thermodynamic system consisting of a hot reservoir at temperature T_H, a cold reservoir at temperature T_C, and a heat engine can be used as an example of how entropy calculations are employed. To be specific, T_H will be assumed to be 600 Kelvins and T_C to be 300 Kelvins. If 5000 kilocalories of heat were to flow directly from the hot reservoir to the cold reservoir, that amount of heat would become totally degraded or unavailable for transformation to some other form. Alternatively, the 5000 kilocalories could flow through a heat engine having an efficiency of 20 percent. Another alternative would be for the heat to flow through a Carnot engine, which in this case would have an efficiency of 50 percent, as can be verified using the concepts discussed above.

If the heat flows directly from the hot to the cold reservoir, the entropy change of the hot reservoir is $-5000/600 = -8.33$ kilocalories per Kelvin (a decrease) as calculated above. The entropy change (an increase) of the cold reservoir was also calculated above as $+5000/300 = 16.67$ kilocalories per Kelvin. Combining the two entropy changes, there is a net entropy increase for the total system of $+16.67 - 8.33 = +8.34$ kilocalories per Kelvin. This number is a measure of the fact that heat has been degraded by passing from the high-temperature reservoir to the low-temperature reservoir; that is, it is less available for transformation.

On the other hand, if the heat flows through the 20-percent-efficient engine, 1000 kilocalories of heat are converted into some other form of energy and 4000 kilocalories of heat are discharged to the cold reservoir. The entropy increase of the cold reservoir is then $4000/300 = +13.33$ kilocalories per Kelvin. The entropy change of the hot reservoir is still -8.33 kilocalories per Kelvin, so the net entropy change of the total system is $+13.33 - 8.33 = 5$ kilocalories per Kelvin, which is less than before; this means that less of the heat energy which could have been transformed has been degraded.

Finally, if the heat flows through the 50-percent-efficient Carnot engine, 2500 kilocalories are converted into some other form of energy and 2500 kilocalories are discharged to the cold reservoir. The entropy increase of the cold reservoir is $+2500/300 = 8.33$ kilocalories per Kelvin. The entropy change of the hot reservoir is still unchanged, and the net entropy change of the total system is $+8.33 - 8.33 = 0$. In other words, for the Carnot engine, all the energy available for transformation is transformed to some form other than heat and is still available for further transformation. Thus in this case the entropy of the total system is not changed, although the entropy of various parts of the system (the two reservoirs) did change. Even in the Carnot case, some energy was "lost" because it was transferred to the low-temperature reservoir, but that energy was already understood to be unavailable for transformation because it had previously been converted into heat within the isolated system. Whenever energy is in the form of heat, we know that some of it is unavailable for transformation and is in a degraded form. Only if the heat is discharged through a Carnot engine is it kept from degrading further in subsequent processes.

It should be noted that in the sample entropy calculations above, the entropy of the total system either increased or was unchanged. The question then arises as to whether there could be any changes in the total system such that there would be a net entropy decrease. In the particular case considered above, this would require that the entropy increase of the low-temperature reservoir be less than 8.33 kilocalories per Kelvin. This would mean that less than 2500 kilocalories would be discharged from the engine, and therefore more heat energy would be converted into another form in the Carnot engine than is allowed by the second law of thermodynamics. But the second law cannot be violated. This is an example of yet another version of the second law: *The entropy of an isolated system can never decrease; it can only increase or remain unchanged.* This statement can be contrasted with the first law of thermodynamics: the energy of an isolated system must remain unchanged—it can neither increase nor decrease. The entropy, on the other hand, may increase: entropy is "semiconserved." The first law of thermodynamics makes it possible to define the internal energy of a system; the second law makes it possible to define the entropy of a system.

Entropy can be looked upon as a substance which has some of the properties that were attributed to caloric: it can "flow" from high temperatures to low temperatures, and in particular it flows from a heat source through a Carnot engine and into a heat sink, causing the heat wheel to rotate and put out work. Whenever the system goes through a reversible process, the total entropy of the system is conserved. In any reversible process in the system, some energy may be transferred within the system. In an irreversible process, some additional entropy is created (rather than lost, as would be the case for caloric). Clausius, in his original conception of entropy, apparently thought in these terms. However, it is more useful to think of entropy as a parameter of a system.

Whenever some other form of energy in a system is converted into heat, the entropy of the system increases. In Joule's paddlewheel experiment, the falling weights start with potential energy determined by their mass and their height above the ground. As they fall, their mechanical potential energy is converted into "heat" added to the water. The total energy of the system consisting of the water and the weights is conserved; however, the entropy of the water is increased (calculated as in the examples above), whereas the entropy of the weights is unchanged (no heat was added or taken from them). As a result, there has been a net entropy increase in the total system because of the transformation of mechanical potential energy into heat energy.

The examples of entropy calculations discussed above and others like them show that it is possible to use the numerical value of the entropy of various systems to characterize mathematically the ways in which the energy is distributed throughout the systems. The flow of heat from a hot part to a cold part of a system is a redistribution of the energy of the system, and as a result the entropy of the system as a whole increases. Of course, in time the entire system comes to thermal equilibrium at some uniform temperature. In that case, the entropy of the system, having increased as heat flowed, has reached a high point. Because all parts of the system are now at the same temperature, no more heat will flow, and thus the entropy of the system is said to be "at a maximum" when the system is in thermal equilibrium.

It is possible to disturb this equilibrium by converting part of the energy of the system into additional heat (for example, some coal or oil in the system can be ignited and burned). This will increase the entropy of the system, marking the fact that the system is now quite different because the chemical energy of the coal or oil is now in the form of energy of molecular motion and therefore less available for transformation to other forms. This "new" heat energy is now ready to flow to other parts of the system. As discussed above, if it flows through a Carnot engine there will be no further entropy increase; if it flows through any other engine or directly to other parts of the system, however, the entropy of the system will increase to a new maximum value.

Entropy Increase and Irreversibility

If some process that causes an increase in entropy of an isolated system takes place, the second law of thermodynamics states that the process can never be reversed so long as the system remains isolated, because such a reversal would decrease the entropy (from its new value), which is forbidden. Once the coal is burned, producing heat and ashes, the process cannot be reversed to make the heat flow back into the ashes and the ashes become coal again. A process can be reversed only if the accompanying entropy change is zero.

A Carnot engine is completely reversible because the total entropy change associated with its operation is zero. Any engine having any amount of friction (and, therefore, transforming mechanical energy into heat) carries out an irreversible process. Even a frictionless engine that does not operate in the Carnot cycle carries out an irreversible process because the total entropy change of the system of reservoirs plus engine is greater than zero. The engine itself may be reversible,* depending on whether it has friction or not, but nevertheless there has been an irreversible change in the total system. As already stated, this irreversible change is associated with the approach to thermal equilibrium of the two reservoirs.

Thus in this sense, the principle of increase of entropy "tells the isolated system which way to go." The system can only go through processes that do not decrease its entropy. One British scientist, Arthur S. Eddington, referred to entropy as "time's arrow" because descriptions made at different times of an isolated system could be placed in the proper time sequence by arranging them in order of increasing entropy.

Entropy as a Parameter of a System

If a system is not isolated but can be acted on by other systems, it may gain or lose energy. If this energy is in the form of heat, it is possible to calculate the entropy change of the system directly from the heat received or lost, provided the temperature of the system can be calculated while the heat is being gained or lost. This can be done using the first law of thermodynamics as an equation, together with the equation of state of the system. The mathematical analysis shows that the entropy of the system can be treated as a physical parameter determining the state of the system, just as temperature and internal energy are physical parameters determining the state of the system. The value of the entropy can be calculated from the temperature and other parameters of the system such as volume, pressure, electrical voltage, or internal energy content. Every system has entropy. Although there are no "entropy meters" to measure the entropy of a system, the entropy can be calculated (or equivalently looked up in tables) if such other parameters as temperature, volume, and mass are known. Similarly, if there were no "temperature meters" (thermometers), it would be possible to calculate the temperature of an object from its other parameters.

*In principle, a heat engine can be reversed, or run "backward," by putting work into it and causing it to draw heat from a low-temperature reservoir and transferring that heat, plus the equivalent heat of the work put in, to a higher-temperature reservoir. The reversed heat engine is called a heat pump. Examples of heat pumps are air conditioners and refrigerators, which are used to cool buildings and food respectively. Heat pumps are also used to heat buildings. A Carnot engine run "backward" is the ideal heat pump. According to the second law of thermodynamics, the operation of a heat pump in an isolated system can never result in a decrease of the entropy of the system as a whole.

Just as temperature and pressure and the other macroscopic parameters, which are parameters describing gross or bulk properties of the system, are ultimately "explained" in terms of a microscopic model (the kinetic-molecular theory) that considers matter to be made up of atoms and molecules in various states of motion and position, so too entropy and the second law of thermodynamics can be "explained" in terms of the kinetic-molecular theory of matter.

Probability and the Microscopic Interpretation of Entropy

As speculated even as early as the times of Francis Bacon, Robert Hooke, and Isaac Newton, the effect of transferring heat to a gas is to increase the microscopic random motions of molecules but with no resulting mass motion. Even "still air" has all its molecules in motion, but with constantly changing directions (because otherwise the air would have an overall mass motion—there would be a wind or breeze). The absolute temperature of a gas can be shown to be proportional to the average random translational kinetic energy per gas molecule.

It is important to make a clear distinction between random and ordered motion. If a bullet is moving through space at a velocity of several hundred miles per hour, the average speeds of its molecules are all the same and in the same direction; their motion is said to be organized. When the bullet has an adiabatic (no heat lost) collision with another object that brings it to a halt, the molecules still have the same average kinetic energy as before, but the motion has now become totally microscopic and randomized. The molecules now are not all going in the same direction, nor are they traveling very far in any one direction, so their net average velocity (as contrasted to their speed) is zero. The kinetic energy of the bullet, which formerly was calculated from its gross overall motion, has been transformed into heat and added to the energy of the microscopic random motions of the molecules, with the result that the temperature of the bullet has increased.

The use of terminology such as "the average random translational kinetic energy per molecule" implies that some molecules are moving faster and some slower than the average, and that there are other forms of kinetic energy such as tumbling, twisting, and vibrations of the different parts of the molecule. In fact, any particular molecule may be moving faster than average at one time and slower than average at another. One may ask what proportion of the molecules is moving only a little faster (or slower) than the average, what proportion is moving much faster than the average, and so on. A graph of the answers to these and similar questions is called a *distribution function* or a *partition function*, because it shows how the total kinetic energy of the gas is distributed or shared among the various molecules.

It should be, in principle, possible to calculate this distribution function from the basic principles of mechanics developed from the ideas of Isaac Newton and his contemporaries and successors. But this would be a very difficult and complicated calculation because the large number of molecules in just one cubic inch of gas (about 200 million million million, or 2×10^{20}, using "scientific notation") at normal temperature and pressure would require that perhaps an equally large number of equations would have to be solved. Moreover, it would be extremely difficult to make the measurements to determine the starting conditions of position and velocity of each molecule required to make the calculations. The next best thing to do is to try to make a statistical calculation—that is, make assumptions as to where "typical" molecules are and with what velocities they are moving. It is necessary to use ideas of probability and chance in these assumptions. Therefore the result of such assumptions is that there will be random deviations of particular individual molecules from the "typical" speeds and direction.

The essential meaning of *random* is summed up in terms such as "unpredictable" or "unknown" or "according to chance." Nevertheless, although the motions of particular individual molecules may be unpredictable, the average motion of the "typical" molecule is predictable. One can, in fact, even predict what a particular individual molecule will be doing (although only with probability) so that a large enough statistical sample of molecules, such as the particular one, will behave on the average according to predictions.

But this poses a dilemma. How can there be unpredictable results if everything is based on Newtonian mechanics, which assumes certainty? It is necessary to make a further hypothesis or assumption that the averages obtained using probabilities are the same as the values that would be obtained if calculations were performed with Newton's laws and then averaged. This assumption, called the ergodic hypothesis, coupled with the laws of Newtonian mechanics, then leads to the result that the distribution function over a period of time will develop in such a way as to be identical to a random probability distribution about an average kinetic energy.

The microscopic model and the idea of distribution functions can be used to "explain" why heat flows from high temperatures to low temperatures. This is particularly illustrated by the example of mixing a hot gas with a cold gas resulting in a transfer of heat from the hot gas to the cold gas, and in obtaining an equilibrium temperature for the mixture. Figure 5-7 shows a schematic representation of a container containing hot (high-temperature) gas molecules on the right and cold (low-temperature) gas molecules on the left. Initially, the boundary between the two halves of the container was made by an adiabatic (perfectly insulating) wall, but this is now completely removed, so the molecules are free to cross over between the two halves of the container. Even though the "hot" molecules will not travel very far between collisions, they will interact with the "cold" molecules closest to them by colliding with them, and as a result the "cold" or slower-moving molecules will gain energy from the collisions, even if they are

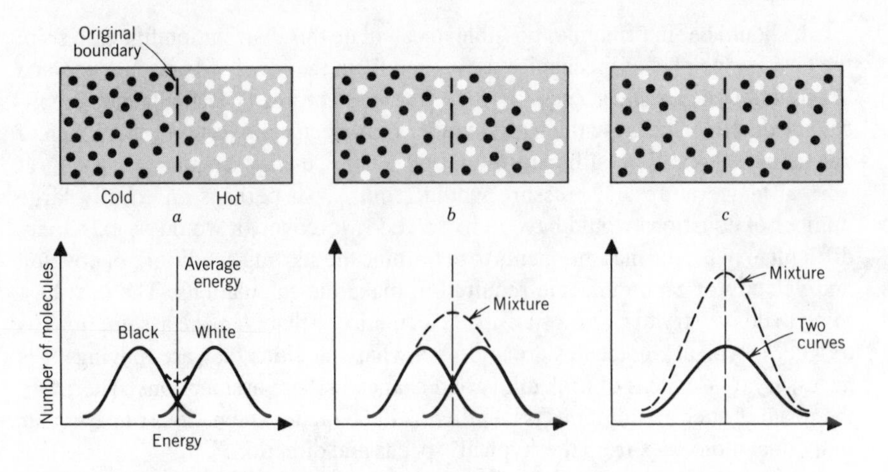

Figure 5-7. Mixing of hot and cold gases. Black molecules originally all on left side and at lower temperature than white molecules, which originally were all on the right side. (a) Original distributions shortly after removing adiabatic wall. (b) Mixing process about half completed. Energy distributions becoming more alike. (c) Mixing process complete. The energy distributions of the two sets of molecules are now identical and merged into one overall equilibrium distribution. The solid curves in the graphs represent the energy distributions for the black and white molecules; the dotted curves, which are the sum of the solid curves, represent the energy distribution for all the molecules.

driven back to their own half of the container. They can then in turn collide with other cold molecules and "share" the energy they gained with the other molecules. In the meantime, some of the cold molecules will diffuse into the right half of the container, increasing their average energy as they collide or interact with the hot molecules. Similarly, some of the hot molecules will diffuse into the left half of the container, losing energy as they collide or interact with the cold molecules. In time, all the hot molecules may interact directly or through a chain of collisions with cold molecules, and similarly all the cold molecules will interact with hot molecules.

As this process continues, the two previously distinct energy-distribution functions will move toward each other, becoming more alike, as shown in Figure 5-7. Eventually they become essentially identical; that is, the two separate distribution functions will become one. The average kinetic energy of translational motion per molecule (corresponding to the absolute temperature) will be the same as the average (weighted according to relative numbers of originally hot and cold molecules) of the two original averages, but now there will be a random distribution of all the molecules about the new average. This new distribution function is the equilibrium distribution function; that is, if the system is not further disturbed, the distribution will not change.

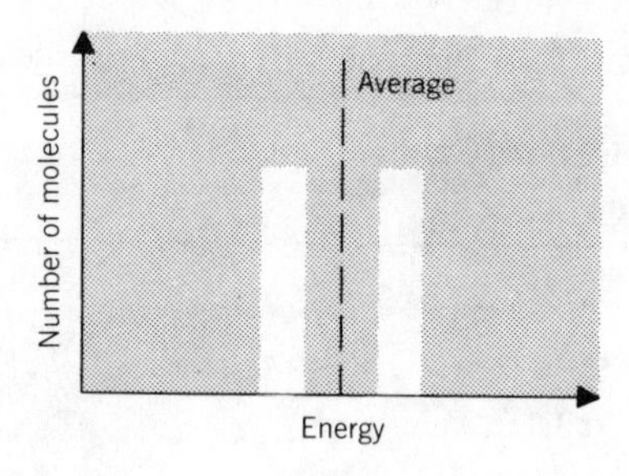

Figure 5-8. An improbable energy distribution.

Moreover, the equilibrium distribution is the most probable distribution function in the following sense: It is possible to imagine many ways in which a given total energy content of a system can be shared among the various molecules of the system. For example, after all the molecules are mixed, half the molecules could have exactly 10 percent more energy than the average and the other half could have exactly 10 percent less energy than the average. Such a distribution function is shown in Figure 5-8. The probability that such a distribution function would occur is very small. It is much more likely that the distribution function will be random about the average value, because there are many more ways of being random (unpredictable) than of being either 10 percent above or 10 percent below the average. (There are cases in which the average value of a random distribution is not the same as the most probable value—see, for example, Figure 7-15b—but these can be regarded as refinements of the discussion above.)

This is exactly analogous to gambling with two dice. If the dice are thrown a great many times, it turns out that the average value of the sum of the the two dice is equal to 7 and the most probable value is also equal to 7. The next most probable values are 8 and 6, whereas the least probable values are 2 and 12. This occurs because, as shown in Table 5-2, there are six different ways to make 7, five different ways to make 8 or 6, but only one to make 2 or 12. Successful gamblers using honest dice are well aware of this.

Because the equilibrium distribution is one of maximum probability in the microscopic picture and the second law of thermodynamics leads to the idea that entropy is maximum at equilibrium, it is reasonable to assume that there is a connection between the entropy of a system and the probability that its particular energy distribution will occur. (It is possible to show that the logarithm of the probability of occurrence of a particular energy distribution for a given system behaves mathematically in the same way as the entropy of the system.) In fact, the microscopic picture coupled with the introduction of concepts of probability makes entropy a very powerful idea. We can explain, for example, why all the molecules in a container do not gather on one side of the container: It is simply more probable for molecules to spread throughout the container, because there are

Table 5-2 Ways in which Two Dice can be Thrown to give a Particular Sum

Sum of Two Dice	Combinations Giving the Sum	Number of Combinations
2	1,1	1
3	1,2; 2,1	2
4	1,3; 2,2; 3,1	3
5	1,4; 2,3; 3,2; 4,1	4
6	1,5; 2,4; 3,3; 4,2; 5,1	5
7	1,6; 2,5; 3,4; 4,3; 5,2; 6,1	6
8	2,6; 3,5; 4,4; 5,3; 6,2	5
9	3,6; 4,5; 5,4; 6,3	4
10	4,6; 5,5; 6,4	3
11	5,6; 6,5	2
12	6,6	1

more ways in which they can be placed throughout the container than ways in which they can be placed alongside one wall. For example, in order for all the molecules to accumulate along one wall, the collisions between them would have always to knock the molecules in the direction of that wall. This is obviously unlikely; the collisions are equally likely to knock molecules in all directions, filling the whole container. Therefore the entropy is greater for the situation in which the molecules are spread out, the equilibrium state. Similarly, if two types of gas molecules are present in a container, it is more likely they will be randomly mixed than segregated into layers (this assumes, of course, that the force of gravity is not so strong that differences in the weight of the gas molecules are significant).

Entropy and Order: Maxwell's Demon

If several truckloads of bricks were dumped onto one location, they would probably fall into a jumbled pile. It is highly unlikely that they would fall into place to make an orderly structure such as a building. This is because there are more ways to make a jumbled pile of bricks than to make a building. In other words, a disorderly arrangement is more probable. (Similarly, it is more likely that a person's living quarteres will be disorganized because there are many more ways to be sloppy than to be neat.)

Returning to the kinetic-molecular theory of matter, when a substance such as water is in a solid state (ice), its molecules are arranged in a very definite regular geometric pattern. As heat is added to the solid, the entropy increases. The molecules, although maintaining almost the same average pattern as before, are moving about much more at any instant of time. The entropy increase is therefore described as an increase of disorder on the molecular level. When the ice melts, the water molecules are moving about even more, and in fact the previous regular geometric pattern has been almost entirely broken, making the system much more disorganized and its entropy therefore much larger. Still, on the average, the molecules are about as close together as before. As more heat is added and the temperature rises well above the melting temperature, the average distance between molecules increases and the range of the momentum values of the molecules increases. When the water evaporates, the molecules become quite separated, and their range of momentum values increases even further. The molecules are now in a very disorganized arrangement and the entropy of the system is therefore much larger. The entropy increase as a result of evaporation is calculated by dividing the latent heat (see Chapter 4) by the thermodynamic temperature.

The ideas of order and disorder (or organization or lack of it) include not only arrangements of molecules in space but also the energy-distribution functions as well. The mixing of hot and cold molecules discussed above in connection with the approach to equilibrium can also be described as a loss of organization in their energy distributions. Initially, half the molecules were clustered in a distribution with a low average temperature (less energy per molecule) and the other half in a distribution with a high average temperature. This represents a discernible grouping or ordering of the molecules—those with low energy and those with high energy. After they have mixed and reached equilibrium there is only one distribution, and there is no longer organization or spatial segregation according to energy.

To sum up, the principle that the entropy of an isolated system cannot decrease simply means that a system by itself will not separate its molecules into groups of different average energy (or spacing or both), because such a separation would represent the acquisition of some higher degree of orderliness or organization—which, statistically, is quite improbable. Furthermore, microscopically, entropy is a measure of the disorder of a system. When the entropy of a system has increased it has become less orderly.

Many schemes have been proposed to circumvent the principle of no entropy decrease for an isolated system. One of the most famous and fanciful of these schemes was considered by the great Scottish theoretical physicist, James Clerk Maxwell (1831–1879). He imagined a box containing a gas of molecules in thermal equilibrium. A wall was mounted across the middle of the box, dividing it into two parts. In the wall there was a trapdoor that was initially open so molecules from both halves of the container could pass through. Maxwell then proposed that there might be a tiny elf sitting by the door who could close or open it very quickly. This elf had the demonic instinct to frustrate the second law of thermodynamics and was therefore called Maxwell's Demon.

The demon wanted to separate the molecules into two groups—"hot" or fast molecules on the right side of the container and "cold" or slow molecules on the left side of the container. The elf would watch all the molecules, paying particularly close attention to those molecules that seemed to be approaching the door. If a fast molecule approached from the right, the demon would quickly slam the door in its face so that the molecule would bounce off the door and head back toward the right side of the box. On the other hand, if a slow molecule approached from the right, the demon would open the door and let the slow molecule pass through to the left side of the box. Fast molecules approaching the door from the left side of the box were allowed to pass through to the right side, whereas slow molecules approaching the door from the left bumped into the closed door and were kept on the left side. Thus after a while the demon would separate the molecules into two groups, one of higher average translational kinetic energy per molecule and therefore of higher temperature on the right, with the other group having lower temperature on the left.

The second law of thermodynamics states that elves and demons exist only in fairy tales, and Maxwell's Demon was only a figment of his imagination, as he himself well knew. The law moreover says that if, instead of a demon, any sort of mechanical or electronic device that could sense molecular speeds were installed to control the operation of the door, such a device will not work unless it receives energy from outside the system. (This conclusion is not obvious from the information presented above, but a detailed analysis shows it to be so.) If energy is received from outside the system, however, the system is no longer isolated.

The second law must be slightly qualified as a statistical law. Because the microscopic picture deals with probabilities, it cannot be said with absolute certainty that a group of molecules cannot spontaneously increase its order by the significant amounts proposed by Maxwell's Demon. But such an occurrence is so rare that it would scarcely be credible if it happened; in any case a second (or third or fourth) occurrence of the event, which would be needed to verify for skeptics that it actually happened, would be so unlikely as to be beyond experience. One should not "hold one's breath" waiting for it to recur.

Cosmological and Philosophical Implications: "Heat Death" of the Universe

As discussed in Chapter 4, Julius Robert Mayer considered that the Earth's energy resources originated in the Sun. For example, the energy stored in coal was derived from the decay of vegetation that had grown because it previously received energy in the form of sunlight. Even the heat energy in the Earth's interior, which is thought to be a result of confined radioactivity in the Earth's core, originally came from the material from which the Earth was formed. But this material, according to one current idea, had its origin in the supernova explosion of some ancient star, of which the entire solar system is a remnant. Generalizing from this postulated history of the Earth and solar system, it is now believed that overall sources of energy in the universe are primarily the stars, which are, of course, quite hot. Depending on their temperature, some of the stars' nuclear energy is released in the form of heat.

If the entire universe is a closed system and there is nothing else, then it might be assumed that there is a fixed amount of energy in this system. This energy is distributed throughout the universe in some manner. The various stars represent very high concentrations of energy, and thus there is order in the universe. It might be reasonably expected that ultimately any particular star will cool and lose its heat. It may go through a rather complicated cycle in doing so. The Sun, for example, might use up all its presently available energy in 10 billion years, according to some estimates. As it cools off, its internal pressure will decrease (according to its equation of state), and as a result of the cohesive force of gravitation the Sun would then collapse, thereby igniting a new series of nuclear reactions, converting some of its nuclear energy into heat and thus raising its temperature again. In time, however, the Sun will radiate away large amounts of energy while cooling to some stable low temperature. Presumably, similar processes should occur in all stars, so that ultimately all the stars in the universe will have cooled off and redistributed their previously available energy.

This process might take a length of time ranging from a thousand million to a million million years (10^9 to 10^{12} years); but if the second law of thermodynamics holds in the closed universe, it must happen, and the universe will reach some equilibrium temperature estimated to be well below 10 Kelvins. All temperature differences will have disappeared; all heat engines will be inoperable; even life itself will be impossible because the energy needed to sustain life will be unavailable. The universe as a whole will have become essentially disorganized, because the equilibrium condition is one of maximum entropy and hence maximum disorder. Moreover, anything that increases the order in any one part of the universe leads to a greater decrease of order in the rest of the universe.

This ultimate end of the universe is called the "heat death of the universe," although the universe will be very cold indeed if this happens. The concept is implicit in the writings of Clausius and of the Austrian physicist Ludwig Boltzmann, but was first explicitly discussed by Lord Kelvin in 1852.

If heat death represents the ultimate state of the universe, then such human ideals as the inevitability of progress and the perfectibility of the human race are delusions. If the ultimate end is chaos, what is the point of scientific endeavor, which seeks order? The idea of heat death has been attacked on philosophical and political grounds. It is anathema to orthodox Marxist theoreticians, who, because of their basic materialistic viewpoint, are very sensitive to implications of science for philosophy. There are other grounds for skepticism, however. The assumption that the universe is a closed system has on occasion been questioned. Furthermore, the assumption that the laws of thermodynamics apply throughout the universe in the same way they apply in our small portion of the universe is an extrapolation from experience based on relatively small systems. Too often, extrapolations of known principles beyond the domain in which they were verified have been failures. We do know, of course, that some physical phenomena in the far reaches of the universe seem to be similar to the same phenomena in the vicinity of the Earth—the analysis of the light from stars both near and far is quite consistent. But not much is actually known about the universe. As yet, there is no direct evidence that the extensions of known theories are not valid; if there were such evidence, efforts would be made to modify the theories appropriately. There is as yet simply not much evidence: observational astrophysics and cosmology are relatively young sciences.

Even if statistical arguments are attempted, however, the extension of various probability assumptions from finite to essentially infinite systems has never actually been justified.

Nevertheless, the ideas of entropy increase in isolated systems, of energy degradation, and of the heat death of the universe have had a significant impact on cultural and philosophic thought since the latter half of the nineteenth century. The increasing degradation of personalities of isolated individuals or groups or societies are major themes in modern thought and literature. The writings of such people as Charles Baudelaire, Sigmund Freud, Henry Adams, Herman Melville, Thomas Pynchon, and John Updike among others show evidence of this. Some economists have also included thermodynamic analogies in their studies, as have individuals concerned with ecology. (See, for example, Zbigniew Lewicki, *The Bang and the Whimper: Apocalypse and Entropy in Literature* [Westport, Conn.: Greenwood Press, 1984]; Rudolf Arnheim, *Entropy and Art* [Berkeley: University of California Press, 1971]; and Nicholas Georgescu-Roegen, *The Entropy Law and the Economic Process* [Cambridge, Mass.: Harvard University Press, 1971]; and their bibliographies.)

Albert Einstein
(Photo courtesy United Press International.)

6

Relativity

The facts are relative, but the law is absolute

The theory of relativity is commonly associated with Albert Einstein, with the atomic bomb and nuclear energy, and with the notion that everything is relative. The idea of relativity did not originate with Einstein, however; nuclear energy is primarily a by-product; and Einstein's original work was as concerned with determining what is absolute as with what is relative. The true significance of Einstein's theory comes from its reexamination of certain metaphysical assumptions and its reassertion of other assumptions, and from its recognition of how physical facts and laws are ascertained. The theory of relativity is sometimes described as a theory about theories, in that other theories must be consistent with the theory of relativity. At the same time, in its applications it demonstrates how abstract and esoteric concepts may have very concrete consequences for everyday life. Among the specific consequences, the recognition of the equivalence of mass and energy and their mutual convertibility led directly to the idea of nuclear energy. Other consequences include insight into the structure of matter, solution of many astronomical problems, revision of cosmological concepts, and deeper insights into the nature of electromagnetism and gravitation. The roots of this theory go back to the Copernican revolution and its accompanying questions about absolute motion, absolute rest, and absolute acceleration.

Galilean—Newtonian Relativity

As discussed in Chapter 3, Newton laid down certain postulates in his *Principia*, stated without proof. The only requirement was that they be consistent with each other. These postulates were necessary in order to have a defined terminology that would serve as a basis for later developments. Among the things postulated were the nature of space and time. Because Newton planned to deduce from the motions of objects the forces acting between them, he had to have a basis for describing motion. Instinctively one thinks of motion through space and time, and thus it becomes necessary to be precise about what is meant by the concepts of space and time.

Newton's definition of space was "Absolute space, in its own nature, without relation to anything external, remains always similar and immovable . . . by another name, space is called extension." Newton also stated that the parts of space cannot be seen or distinguished from each other, so that instead of using absolute space and motion it is necessary to use relative space and motion (further discussed below).

He defined time this way: "Absolute, true, and mathematical time, of itself, and from its own nature, flows equably without relation to anything external, and by another name is called duration." There is a difference between objective time and psychological time. We all know that time flies on some occasions and drags on others, depending on circumstances. Newton recognized that, as a practical matter, objective time is measured by motion. For example, during the day the passage of time can be measured by the apparent movement of the Sun in the sky. Even though Newton recognized that it is relative space and time that are measured, he emphasized that in his system there were the underlying concepts given above. There are, of course, ambiguities in Newton's definitions. For example, to say that time flows equably or smoothly means that there has to be an independent way of determining what is equable or smooth. But these words themselves involve a knowledge of time.

Because Newton recognized that relative position, relative motion, and relative time are usually measured, we must discuss what is meant by relative measurements. If it is stated that a certain classroom in Kent, Ohio, is 40 miles due south of Cleveland, Ohio, then its position is given relative to Cleveland. That same classroom can be described as being 10 miles east of Akron, Ohio. Similarily, an official proclamation might be dated 5747 years after the Creation, or 1986 years after the birth of Christ, or in the two-hundred eleventh year of the independence of the United States, depending upon the reference date from which time is measured.

The location of the classroom was not complete as given above; it is necessary to state on what floor of a particular building the classroom is to be found. In fact, more commonly, locations on Earth are given by latitude, longitude, and altitude above sea level. There are any number of different ways in which location can be specified, but in all cases, three coordinates are necessary to locate a particular point in space, because space is three-dimensional. A point in space is a location on a three-dimensional map called a *frame of reference*. The actual values of the coordinates will depend upon the choice of the origin (the point relative to which all others are measured) of the frame of reference; it can be in Cleveland, Ohio, or Topeka, Kansas, or Greenwich, England, or Moscow, USSR, or Quito, Ecuador. For that matter, the origin does not have to be on the Earth. It can be at the center of the Sun or even in the middle of our galaxy. The actual numbers used for the coordinates depend upon the origin of the frame of reference and also on whether distances are measured in miles or feet or kilometers or degrees.

Once a frame of reference is established, motion is measured with respect to the frame of reference by specifying how the coordinates of a particular object change as time goes on. The velocity of an object is specified relative to a particular frame of reference, and the actual values of velocity may depend on the choice of the frame of reference, because the frame itself may be in motion. For example, a student sitting in a chair in a lecture room has zero velocity relative to any fixed point on the Earth, but relative to the Sun that student is moving with a velocity of 67,000 miles per hour because the Earth (on which the student sits) is traveling at a speed of 67,000 miles per hour relative to the Sun. A rocket traveling at 1000 mph relative to the surface of the Earth might be traveling at 68,000 mph relative to the Sun—or 66,000 mph or some intermediate velocity, depending on its relative direction of motion. Because velocity is a vector quantity (see Chapter 3), the direction of the velocity relative to the Sun will also depend on the position of the Earth in its orbit and the relative rotation of the Earth about its axis. Because the Sun itself is in motion relative to the center of the galaxy, the velocity of the rocket relative to the galaxy has yet another value. The galaxy itself is in motion relative to other galaxies.

Thus both coordinates and velocities are relative quantities. All quantities derived from coordinates and velocities (see Chapter 3) must also be relative: momentum, kinetic energy, and potential energy, for example.

In view of Newton's definition of absolute space and time, it is reasonable to ask what the true velocity of the Earth is as it travels through space—not its velocity relative to the Sun or the galaxy but relative to absolute space. To measure that velocity requires a reference point that is fixed in absolute space. One suggestion is to take the midpoint of the average position of the fixed stars, but what proof is there that the stars are fixed in space? These questions are clearly related to the problem that concerned Ptolemy, Copernicus, Galileo, Kepler, and their contemporaries: Is it the Earth that is moving, or the stars, or the Sun, or all of them? Galileo asserted that it is impossible to tell from an experiment done on the Earth whether the Earth is moving in a true absolute sense.

As Galileo pointed out, if a ship is moving in a harbor, an object dropped from the mast of the ship falls straight down and hits the deck at the foot of the mast of the ship, as seen on the ship. But as seen from the shore, the object cannot fall straight down—it must undergo projectile motion. It maintains its forward motion because the ship is carrying it along, and simultaneously it falls toward the Earth. This results in a parabolic path, as seen from the shore. Nevertheless, the object will still hit the deck at the foot of the ship's mast, because the ship was moving along and keeping pace with the forward motion of the object. The sailors on the ship cannot tell whether the ship is moving by watching the falling object, because it hits in the same place on the ship regardless of the ship's motion.

Only by looking at the shore can they tell that there is relative motion between the ship and the shore. Of course, they say that the ship is moving; it is hard to believe that the ship is standing still and the shore is moving. Similarily, Galileo (like Copernicus before him) found it hard to believe that the Earth was standing still and all the stars were moving. But he recognized that he had not proved which is actually moving, or how fast.

In fact, the law of inertia—Newton's first law of motion—states that so far as forces are concerned, all reference frames which move at constant velocities with respect to each other are equivalent. It is impossible to determine which one of these reference frames is actually at rest in absolute space or which one is going faster than the others. An object moving at constant velocity in one reference frame is also moving at a different constant velocity in another reference frame that moves at constant velocity with respect to the original reference frame. The object is also at rest in yet another reference frame.

For example, a man throws a ball at a speed of 60 miles per hour from the rear toward the front of a ship that is itself traveling eastward at 30 mph. Relative to the ship, the ball is moving at a velocity of 60 mph eastward; relative to the ocean, the ball is moving at a velocity of 90 mph eastward. In the absence of external forces (gravity and wind resistance), the ball will continue to move at those same velocities. Moreover, if a helicopter is also flying eastward at a speed of 90 mph relative to the ocean, the ball has a velocity of zero relative to the helicopter: it is at rest. We cannot say that the ball is actually moving at 90 mph because the Earth itself is in motion.

All reference frames in which Newton's laws of motion, in particular his first law, are valid are called *inertial reference frames*. These reference frames may be in motion relative to each other, but their relative motion is at constant velocity. It will be worthwhile here to contrast an inertial reference frame with a noninertial reference frame. An automobile traveling at a steady 50 mph in the same direction can be used as an inertial reference frame. If the driver of the automobile suddenly applies the brakes, while it is decelerating the automobile becomes a noninertial or accelerated reference frame. Objects within the automobile suddenly begin accelerating (with respect to the automobile) even though no force is applied. A passenger in the front seat who is not wearing a seat belt is accelerated from rest and crashes through the windshield of the car. The law of inertia (Newton's first law) has been violated in this reference frame (because the passenger accelerated with no force acting upon him) and so it is called a noninertial reference frame. A rotating reference frame is also a noninertial reference frame. In making calculations, noninertial reference frames are often rejected or discarded, if at all possible, because Newton's laws of motion are not valid in such frames.*

Because all dynamical quantities are measured in terms of space and time, the measurements that are made may depend upon the reference frame used in carrying out the measurements. The velocity of the ball discussed above depends upon the reference frame from which the ball is examined. Similarly, the kinetic energy and the potential energy of the ball will depend upon the reference frame. Of course, it should be possible to determine the physical properties of the ball in any reference frame. Given the velocity of the ball in one reference frame (such as with respect to the ship) it should be possible to calculate its velocity, or any other dynamical quantity, relative to the helicopter.

The equations that make it possible to carry out such calculations are called transformation equations. In effect, transformation equations are like a dictionary that permits "words" spoken in one reference frame to be "translated" into another reference frame. In using transformation equations, the coordinates and velocities measured in one frame of reference are used to calculate the coordinates and velocities that would be measured in another reference frame. (Because frames of reference can be in motion relative to each other, the transformation equations also involve time, a factor that will be discussed further below.)

As already mentioned, many quantities or measurements will be different in different frames of reference: position, velocity, momentum, kinetic energy, potential energy. On the other hand, some quantities do not change, regardless of which inertial frame of reference is used. Such quantities are called *invariant* or constant. In Newtonian physics, examples of such invariant quantities are mass, acceleration, force, electric charge, time, the distance between two points (length), temperature, energy differences, and voltage differences. When these quantities are measured in one frame of reference and their values are calculated in another, using the appropriate transformation equations, the quantities have exactly the same values as in the first frame of reference. If the quantities in the second frame of reference are measured rather than calculated, they will again be exactly the same as measured in the first frame of reference.

But more than quantities are invariant. Newton's laws of motion are invariant from one inertial frame of reference to another. This means that the mathematical form of Newton's laws of motion is the same in all inertial frames of reference. The law of conservation of energy is the same in all inertial frames of reference, even though the total amount of energy will be different in the different inertial

*Strictly speaking, the Earth is a noninertial reference frame because it does not travel in a straight line and also because it spins on its axis. The effect of the resulting acceleration is very small compared with the effect of the force of gravity at the surface of the Earth, however, so for many purposes the error made by assuming the Earth to be an inertial reference frame is not too large. Nevertheless, it is the noninertial effects of the Earth's spin that cause cyclones and tornadoes.

frames. Similarly, the law of conservation of momentum is the same in all inertial frames of reference, even though the total momentum is different in different inertial frames.

Galilean–Newtonian relativity is the name given to the aspects of relative and invariant quantities and relationships and transformations between inertial reference frames, as discussed so far. These relationships and transformation equations depend in a very fundamental way on the concepts of time and space.

Electromagnetism and Relative Motion

Two of the great triumphs of nineteenth-century physics were the development of the concept of energy with all of its implications and the development of a unified theory of electricity and magnetism. The technological fruits of these two theories have shaped our material culture, and their scientific consequences have been equally pervasive.

As the theory of electromagnetism developed, it raised once again the possibility of determining the velocity of the Earth in absolute space. Whereas according to Newtonian mechanics there was no experimental way of determining whether the Earth was "really" moving in an absolute sense, at least from a mechanical experiment, electromagnetic theory suggested a number of experiments that could be carried out to determine the "true" motion of the Earth.

Electromagnetic theory is also important for understanding the structure of matter, a subject of importance in the following two chapters, so it is worth discussing certain aspects of electricity and magnetism here.

It is particularly easy to generate static electricity when the air is dry, as in the winter in cold climates, by means of a phenomenon known as the triboelectric effect. If one removes an article of clothing such as a pullover sweater, or brushes long dry hair, or strokes a cat, or walks across an untreated wool carpet, one feels a tingle, hears crackling sounds, sees flashes of light, or observes static cling of items of clothing. All these effects, and many more, are due to the generation of static electricity by rubbing two or more objects together. This effect was known more than twenty-six hundred years ago by Thales, the Ionian philosopher who first recognized the value of studying mathematics.

Static electricity is generated by the transfer of *electric charge* from the surface of one object to another object. In time it was recognized that there are two kinds of electric charge, one called positive (+) and the other negative (−), and that it is possible to separate positive and negative charges from each other. Positive charges repel positive charges and negative charges repel negative charges, but positive charges and negative charges attract each other. The mathematical formula for calculating the strength of the forces between two charges is called Coulomb's law and is similar to the universal law of gravitation: The force is proportional to the product of the magnitude of the charges and inversely proportional to the square of the distance between them. The magnitude or amount of an electric charge is designated by the symbol q. The algebraic formula for the force between the two charges Q and q is $F = kQq/r^2$, where k is a proportionality constant and r is the distance between the charges. Experimentally, it is found that charges come in multiples of one definite size, called the charge of an electron, which is designated by the letter e. (The most recent theory of the structure of subnuclear particles proposes that charges of two-thirds and one-third of e also exist, but such charges have not been detected in a free state; see Chapter 8.)

Although Coulomb's law makes it possible to calculate the force between two electric charges, it does not explain how that force is transmitted from one charge to another. This is the same question that was noted in Chapter 3 in connection with Newton's law of universal gravitation: How do objects manage to "reach" across empty space and exert forces on each other? (This question is discussed later in this chapter and in somewhat more detail in Chapter 8.) One way of dealing with questions such as this, about which very little is known, is to give the answer a name. In this case, the means by which one electric charge exerts a force on another electric charge is called the electric field.

Electric fields are said to be inherent in the very nature of electric charges. An electric charge q is said to have an electric field, designated by the symbol E, associated with it. E is a vector quantity (see Chapter 3) and extends over a fairly large region of space, although its magnitude becomes smaller at increasingly greater distances from the electric charge q. Figure 6-la gives a schematic representation of the electric field from a concentrated electric charge (called a point charge). The arrows indicate the direction of the field; the magnitude of the field depends upon how close together the arrows are. Because the arrows are directed radially, as shown in the figure, they are closer together near the charge and farther apart far from the charge, showing that the magnitude of the field is greater close to the charge and less far from the charge.

If there are several electric charges present in a given region of space, the combined effect of the electric fields is calculated by vectorially adding their individual electric fields. Figure 6-lb shows the resulting electric field from a pair of positive and negative charges of equal magnitude, and Figure 6-lc shows the electric field resulting from a pair of positive charges. Figure 6-ld shows the resulting electric field when sheets of positive and negative charges are placed opposite each other. If still another electric charge is brought into the region of space where an electric field has been established, then that other charge will be subjected to a force proportional to the combined electric fields of the charges already there, and in the direction of the combined fields (for a positive charge; in the opposite direction for a negative charge). The electric field can be considered a useful way of calculating the forces exerted by a collection of electric charges on another charge. As indicated below, however, the electric-field concept has more profound uses.

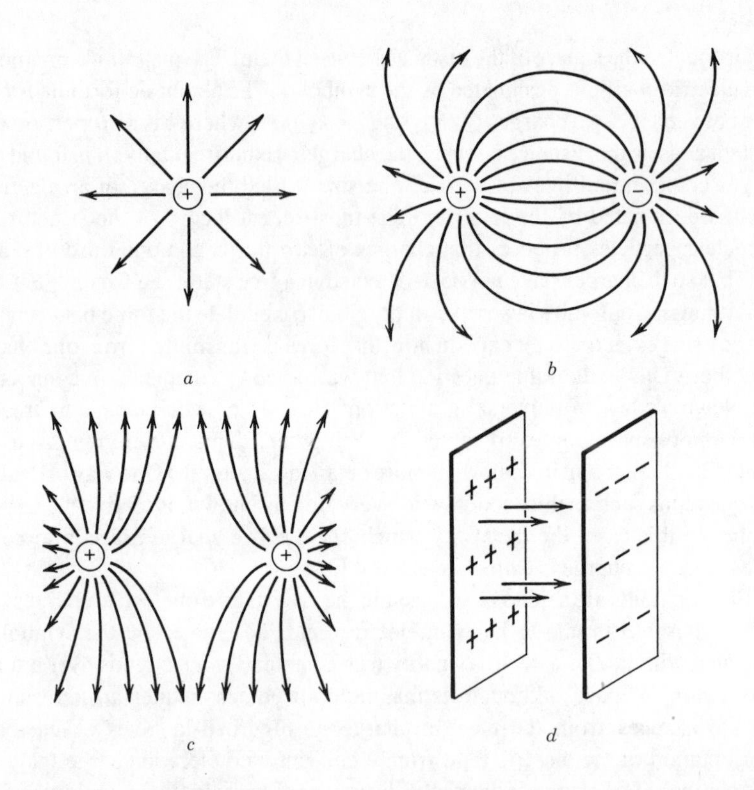

Figure 6-1. Representation of the electric field. The plus and minus signs represent positive and negative charges, the arrowed lines represent the electric field. (a) Isolated positive charge. (b) Equal + and − charges. (c) Equal + charges. (d) Sheets of + and − charges.

Magnetic effects have also been known from ancient times. In many ways magnetic effects are similar to electric effects. Magnetic charges are given a different name than electric charges; they are called magnetic poles. There are north poles (sometimes called north-seeking poles) and south (or south-seeking) poles. There are also magnetic fields. However, it is impossible (at least to date) to separate north poles and south poles as distinct entities in the same way it is possible to separate positive and negative charges from each other. Moreover, all presently known magnetic effects can be attributed to the motion of electric charges.

When one or more electric charges are in motion, the moving charges constitute an electric current. (Electric currents are measured in units called amperes.) Associated with an electric current is a magnetic field, designated by the symbol B. The relationship between B and its associated electric current is quite different from that between E and q, as shown in Figure 6-2. If, for example, the electric current is along a wire (because of an electric field directed along the

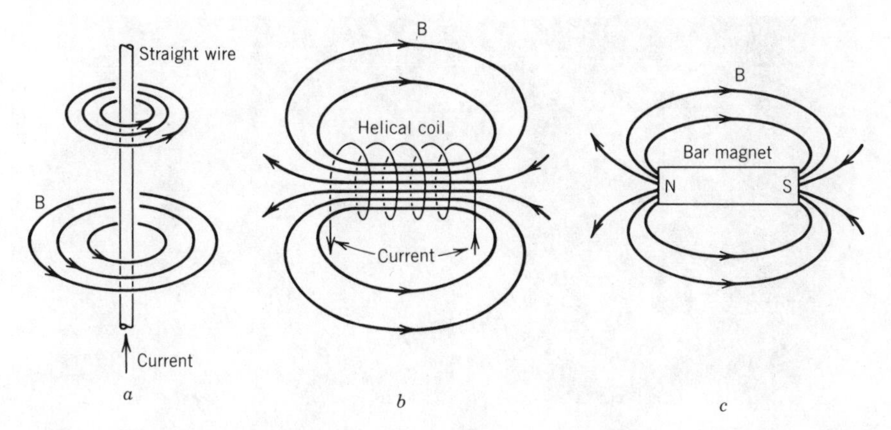

Figure 6-2. Magnetic fields. (a) From upward current in long straight wire. (b) From current in a helical coil of wire. (c) From a bar magnet.

wire), the associated magnetic field is directed in a ringlike pattern around the wire (Figure 6-2a). The magnitude of the magnetic field decreases with increasing distance from the wire. If the wire is given a tight helical shape (Figure 6-2b), the magnetic field inside the helix is directed along the axis of the helix, flaring out at the ends of the helix. The pattern of the magnetic field exterior to the helix is identical to that of a bar magnet (Figure 6-2c), including the location of north and south poles.

In general, magnetic effects in matter result from the presence of electrical currents associated with the dynamics of atomic and molecular structure. In other words, magnetic effects are due to the motion of electric charges, and thus the very existence of magnetic effects is an aspect of relativity. Magnetic effects require only relative motion of electric charges.

Magnetic fields are similar in many ways to electric fields: Magnetic fields due to different currents or moving charges have combined effects that are calculated by adding vectorially the magnetic fields of the individual currents or moving charges. Also, a magnetic field will exert a force on any other current or moving charge that may be present.

Magnetic fields raise the same problem as electric fields. The magnetic field is the mechanism by which a moving electric charge (or an electric current) exerts a force on another moving electric charge. Formulas were developed for calculating the force, but without a clear explanation about how one moving charge "reaches" across empty space to exert a force on another moving charge. (This question will also be addressed in Chapter 8.) As in the case of the electric field, however, the magnetic field is more than just a useful way of calculating the force. In fact, magnetic fields, like electric fields, can even be made visible, as shown in Figure 6-3, which illustrates the use of iron filings to reveal the magnetic fields of wires and bar magnets. Calculating the force is somewhat more complicated than

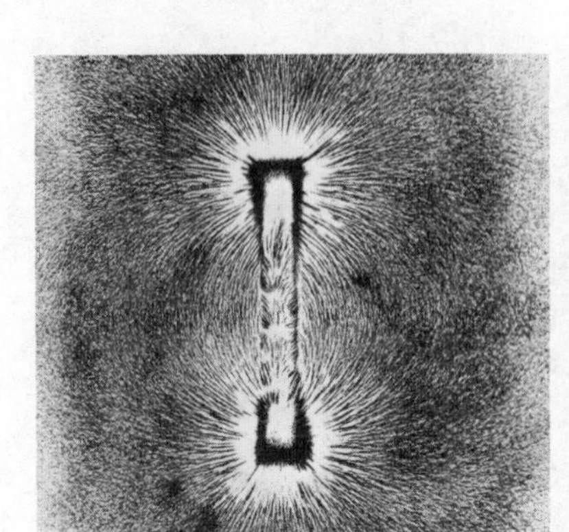

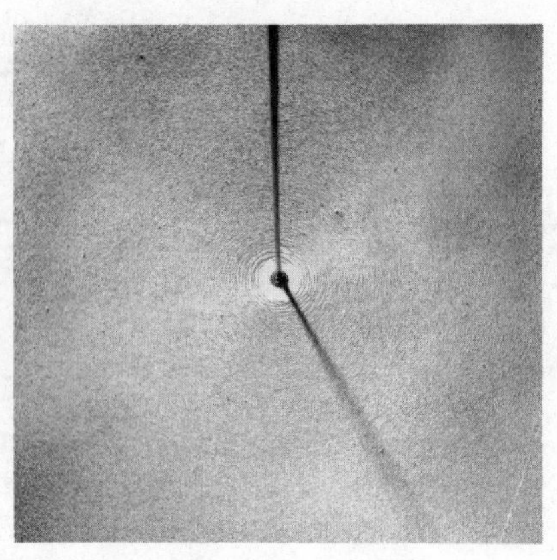

Figure 6-3. Photographs of alignment of iron filings in the presence of a magnetic fields. (a) In the field of a bar magnet. (Kathy Bendo) (b) In the field of a current-carrying wire. (Educational Services, Inc.)

calculating the force involved with an electric field. The force depends not only on the magnitude of the magnetic field and of the moving charge but also on the velocity of the motion, including both speed and direction. The force is also perpendicular both to the original magnetic field and to the original direction of motion of the charge.

Because the force exerted by the magnetic field depends upon velocity, relativistic effects are also present. This allows the possibility for detecting absolute motion and also for leading physics into a paradox: If the velocity depends upon the frame of reference and force depends upon velocity, then force should depend upon the frame of reference. But according to Galilean–Newtonian relativity force should be an invariant and should not depend upon the frame of reference!

Figure 6-4 shows a symbolic scheme for displaying the relationship between electric charges and electric fields and between magnetic charges and magnetic fields. At the apex of the triangular cluster of symbols is q and in the lower right corner is E. Arrow 1 indicates that associated with (or created by) an electric charge is an electric field. Arrow 2 indicates that an electric field will exert a force on any other electric charge present. In fact, if an electric charge has a force exerted on it, then it must be true that an electric field is present. Arrow 3 indicates that associated with a moving electric charge (an electric current) is a magnetic field, B. Arrow 4 indicates that a magnetic field will exert a force on any other moving electric charge that may be present. Because motion is relative, a moving magnetic field will exert a force on a stationary charge.

Even a changing magnetic field exerts a force on a stationary charge. However, as already discussed in connection with Arrow 2, whenever an electric charge feels a force, then there must be an electric field present. Thus, in effect, a

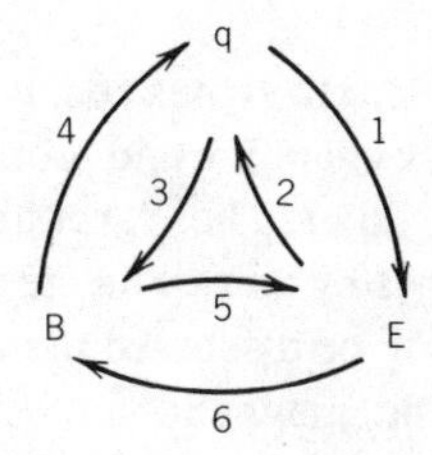

Figure 6-4. Triangle diagram showing relationships between q, E, and B. The arrows shown are not vectors but represent relationships discussed in the text. (By permission from Kenneth W. Ford, *Basic Physics*. New York: John Wiley & Sons, 1968.)

changing magnetic field creates (or is equivalent to) an electric field. This is shown in Figure 6-4 by Arrow 5 between B and E. The reciprocal relationship is also true: A changing electric field will create a magnetic field.

The six relationships represented by the arrows in Figure 6-4 are involved in the technological applications of electromagnetism. For example, electric fields are used to create the beam of electrons that light up the picture tube on a television receiver. Magnetic fields due to electric currents are used in relays and electromagnets. Magnetic forces on electric currents are the basic mechanism at work in electric motors and generators, and magnetic forces on moving electric charges are used to deflect the electron beam to make rasters on television picture tubes. The creation of electric fields by changing magnetic fields represents the essential action of electric transformers.

All of these possibilities materialized in the explosive development of science and technology in the nineteenth century, but the crowning achievement (from the standpoint of pure science, at least) was the realization by James Clerk Maxwell, the creator of Maxwell's Demon, that the phenomena represented by Arrows 5 and 6 in Figure 6-4 are significant for understanding the nature of light. Arrow 5 indicates that a changing magnetic field creates an electric field. The magnitude of the resulting electric field depends on the rate at which the magnetic field changes. If that rate of change itself is steady, the resulting electric field will be steady. If, on the other hand, the rate of change of the magnetic field is not steady, the resulting electric field will not be steady—it will be changing. But Arrow 6 represents the fact that a changing electric field will in turn create a magnetic field. Furthermore, a changing electric field that is changing at a nonsteady rate will create a nonsteady (changing) magnetic field. A continuous chain of creation is possible: A nonsteadily changing magnetic field will create a nonsteadily changing electric field, which creates a nonsteadily changing magnetic field, which creates a nonsteadily changing electric field, and so on.

Maxwell was able to show that these changing electric and magnetic fields that are constantly recreating each other are also propagating (spreading out) through space at a definite speed, which he was able to calculate: 186,000 miles per second—the speed of light! Maxwell thus concluded in 1864 that light could be explained as arising from rapidly changing and propagating electric and magnetic fields closely coupled with each other. These are called electromagnetic fields. In fact, light represents only a very tiny part of the general phenomena of propagating electromagnetic fields, as will be discussed in more detail below.

In his study of electric and magnetic forces Maxwell, like most of his contemporaries, was concerned with the nagging fundamental question (which also applies to the gravitational force): Just what is the mechanism by which one electrical charge (or one mass) reaches across empty space and exerts a force on another electrical charge (or mass)? Newton said—as noted in Chapter 3—"I make no hypotheses." One possible answer is "Through its electric (or gravitational) field." Such an answer was not very satisfactory to nineteenth-century scientists since it does not reveal what an electric field is.

Michael Faraday proposed an analogy between the arrows used to show the electric fields (see Figure 6-1) and rubber bands: The arrows start on a positive charge and end on a negative charge. They are under tension like stretched rubber bands and therefore exert a force of attraction along their lengths between the positive and negative charges (Figure 6-1b). In addition to contracting along their length, the rubber bands tend to push apart from each other, thereby explaining why two charges of the same sign repel each other (Figure 6-1c) and why the pattern of spreading of the bands is as shown in the various parts of Figure 6-1. These bands are called field lines or, somewhat more loosely, lines of force. Similarly, the magnetic field can be described in terms of magnetic field lines connecting magnetic poles and the gravitational field in terms of gravitational field lines connecting masses.

It is possible to use the idea of field lines to actually work out the mathematical formulas for the forces between electric charges, but it is rather difficult to believe that they have objective reality. Nevertheless, in the spirit of Plato's Allegory of the Cave, one might attempt to justify them as representing the "true reality" behind the appearances. In the same way, the followers of the Ptolemaic theory conceived of the heavenly spheres as discussed in Chapter 2. Of course, it is rather difficult to see how various objects could move about without getting all their electric, magnetic, and gravitational field lines tangled.

A more sophisticated and elegant way of approaching the problem of "action at a distance" was suggested. It might be asserted (as Aristotle had done) that there is no such thing as empty space, but rather all space is filled with a substance called the ether, a special sort of substance having no mass whatsoever. (Aristotle's ether was present only in the regions above the lunar sphere. By the nineteenth century the ether was considered to be everywhere.) If a material body, or an electric charge, or a magnetic pole is inserted in the ether, the ether has to be distorted or squeezed in some fashion to make room for whatever is inserted. The electric field lines represent the distortions of the ether resulting from the insertion of an electric charge. The propagating electromagnetic field that is light represents a fluctuation of the distortions present in the ether.

By the early part of the nineteenth century the mathematical theory of elasticity had been thoroughly worked out and verified. Maxwell was able to adapt the theory of elastic solids to this hypothetical electric ether and to represent both electric and magnetic fields as ether distortions. Maxwell described light as a wave of distortions propagating through the ether. In this sense, electromagnetic waves are just like sound waves in air, or in water, or in any other medium. The mathematical theory of such waves had been worked out in terms of properties of the medium. Electromagnetic waves have special properties because the medium for these waves, the ether, is different from air or water. (One of the differences is that waves in air, water, or other fluids are longitudinal whereas waves in a solid can be transverse as well as longitudinal.) One of the significant results of Maxwell's work was that he was able to show that it is not necessary to have one ether for electrical effects, another ether for magnetic effects, and yet a third ether for light. One ether is sufficient—the electromagnetic ether—and different kinds of electrical, magnetic, and electromagnetic (light) effects are simply different types of distortions and combinations of distortions in the same ether. (Of course, he had not included gravitational effects, so conceivably there might be a separate ether for gravity.)

As already noted, in 1864 Maxwell showed that it is possible to develop waves in the electromagnetic field and that these waves should travel with the speed of light. Actually, less than fifty years earlier there had been a lively controversy about the nature of light. There had even been some thought in antiquity that light originated in the eyes of a person looking at an object. Indeed, the expression "cast your eyes upon this" carries with it the implication that the eyes send out a beam that searches out the object to be seen. Even the comic-strip character Superman sees in this way, except that his vision is particularly penetrating because he sends out X rays so that he can see through walls.

Isaac Newton did not consider that the eyes sent out a beam of light, but he did believe that light consists of a stream of corpuscles (little bullets) that makes objects visible by bouncing off the objects into the eyes of the beholder. One reason for his belief was the fact that light seems to travel in straight lines, as evidenced by the sharpness of shadows.

Some of Newton's contemporaries, however, pointed to evidence that light is some sort of disturbance transmitted by a wave motion, just as sound is an air disturbance transmitted by a wave motion or as a disturbance or distortion of the surface of a body of water is transmitted by a wave motion. In time, the experimental evidence in favor of the wave theory of light became so overwhelming that it was universally accepted.

One of the characteristics by which a wave is described is the distance between successive crests of the wave, the wavelength. Wavelengths for visible water waves may vary from a few inches to tens or hundreds of feet. The wavelengths of audible sound waves in air range from a few inches to a few feet. The wavelengths of the light to which our eyes are sensitive, on the other hand, are quite short— roughly 1/20,000 of an inch—a fact known by the beginning of the nineteenth century. It is because of these extremely short wavelengths that light can be formed into "beams" that seem to travel in straight lines and give rise to sharp shadows. Figure 6-5 demonstrates the effect of wavelength on "shadow" formation for water waves. In Figure 6-5b, where the wavelength is shorter, a sharper "shadow" is apparent.

The wave nature of light is also responsible for the appearance of colors in the light reflected from soap bubbles or from oil films on streets just after a rain. The colors result from an effect called interference, discussed further in Chapter 7. The wave nature of light is also manifested in an effect called diffraction, which is responsible for the colors seen from "diffraction jewelry" or when light is reflected at a glancing angle from video discs or phonograph discs, for example those used for 33-1/3 or 16-2/3 rpm record players.

Another characteristic property of waves is their frequency, the number of wave crests per second that will pass by a given point as the waves spread out. The light we see has a frequency of roughly 500 million million wave crests per second. Multiplying the frequency of a wave by its wavelength gives its speed: 186,000 miles per second for light.

Until Maxwell's work, it was thought that light waves were transmitted through a special ether. A wave must travel in some medium; without any water there cannot be a water wave, and without air there cannot be a sound wave, as is demonstrated by the fact that sound waves do not exist in vacuum. For light

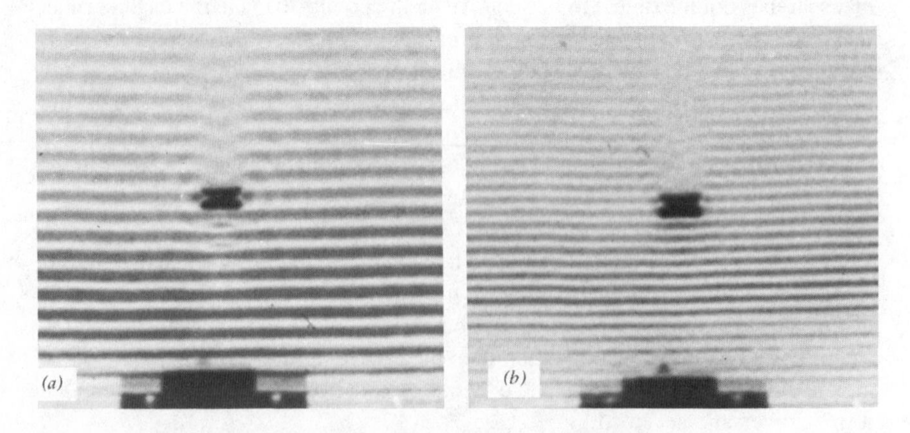

Figure 6-5. Photograph of water waves in a ripple tank (Reproduced by permission of Educational Development Center, Newton, Mass.)

waves, therefore, it was assumed that there is a special ether, called the luminiferous ether, which fills all space. As already indicated, Maxwell was able to show that the luminiferous ether, the electric ether, and the magnetic ether should be all one and the same, and that light waves represent a disturbance of this ether.

One of the important results of Maxwell's electromagnetic theory was the realization that light waves represent only a very small portion of the electromagnetic waves that can be generated. For example, radio waves can be generated, as was demonstrated in 1886 by Heinrich Hertz, a German physicist. Electromagnetic waves are generated when an electric charge is caused to vibrate (oscillate). In effect, the electric charge "waves" its electric field much as a hand waves a long whip. Of course, the waving electric field is more like a bundle of rubber bands, to use Faraday's analogy; but this gives the nonsteadily changing electric field needed to create the electromagnetic wave, as described above. Hertz was able to generate oscillating electric currents at frequencies of a few hundred million vibrations per second. The wavelength of these waves was two to three feet, depending on the frequency. Subsequently, an Italian engineer, Guglielmo Marconi, invented the wireless telegraph to send messages by radio waves. The first successful transatlantic radio messages were carried by waves having a frequency of 200,000 vibrations per second and a wavelength of almost a mile!

The entire range of electromagnetic waves is called the electromagnetic spectrum; its extent is indicated in Table 6-1. All the electromagnetic waves listed in the table travel in empty space at the speed of light, 186,000 miles per second.* The differences among them are due entirely to their wavelengths or frequencies. Waves with short wavelengths have high frequencies, and vice versa. When wavelength and frequency are multiplied together, the result is constant and is equal to the speed of the waves. Note that in Table 6-1 wavelengths are given in meters, as is usually done for radio waves.

An electromagnetic wave can be described as a combination of pulsating electric and magnetic fields coupled to each other. Figure 6-6 shows how the electric and magnetic fields vary in an electromagnetic wave traveling in the direction of the heavy arrow. In the figure, the electric field vectors E are shown as varying in magnitude along the direction of travel but being confined to the vertical plane. The magnetic field vectors B also vary along the direction of travel and are confined to the horizontal plane. When E and B are confined as shown, the electromagnetic wave is said to be plane-polarized. It should be noted that the pulsating E and B vectors are always perpendicular to each other and to the direction of travel; this perpendicularity is true of all electromagnetic waves. Waves perpendicular to the direction of travel are called transverse waves.

* In a material medium such as glass or water, the speed of light in less than in empty space. The ratio of the speed of light in empty space to its speed in the material medium is called the index of refraction of the medium. Most often in a material medium the speed of the electromagnetic waves (and hence the index of refraction) is no longer constant but is different for various wavelengths, a phenomenon called dispersion.

Table 6-1 The Electromagnetic Spectrum

Kind of Wave	Frequency in Hertz*	Wavelength in Meters*
Ordinary household alternating currency	60	5×10^6 (3100 miles)
A.M. Radio	10^6	300
Television	10^9	0.3
Microwaves/radar	10^{11}	0.003
Infrared	10^{13}	0.00003
Visible light	5×10^{14}	6×10^{-7}
X rays	3×10^{18}	10^{-10}

*Hertz means cycles or waves/second; 1 meter equals 39.37 inches; 10^6 means a one followed by six zeroes, that is, one million; similarly, 10^{11} means a one followed by 11 zeroes; 10^{-7} means one divided by 10^7. The product of frequency by wavelength is 3×10^8, which is the speed in empty space of electromagnetic waves in meters/second.

By contrast, the disturbance carried by sound waves traveling in air is in the form of a pulsating density of the air. This pulsation takes place in the direction of travel of the wave and is therefore called a longitudinal wave. When a sound wave travels in a solid medium, the resulting waves can be transverse as well as longitudinal. There are no longitudinal pulsations for electromagnetic waves.

In any wave energy is associated with the pulsations. Transmission of the pulsations by the wave results in the "transport" of energy. An electromagnetic wave transports electromagnetic energy. The rate at which it transports energy is related to the product obtained by multiplying E by B (using special rules for multiplying vectors together). The immediate source of the energy transported by the electromagnetic wave is the kinetic energy of the vibrating electric charge that is waving its electric field. The generation of electromagnetic energy as described above is a process for converting kinetic energy into radiant energy (see Chapter 4).

With the development of electromagnetic theory and its concomitant electromagnetic ether, the question arose as to what happens to the ether when an object such as the Earth moves through it. Does the Earth move along without disturbing the ether, or does the Earth drag the ether along with it? If the Earth does not disturb the ether as it moves along, then it might be possible to use the ether as a frame of reference. It would then be possible to calculate the "true" force

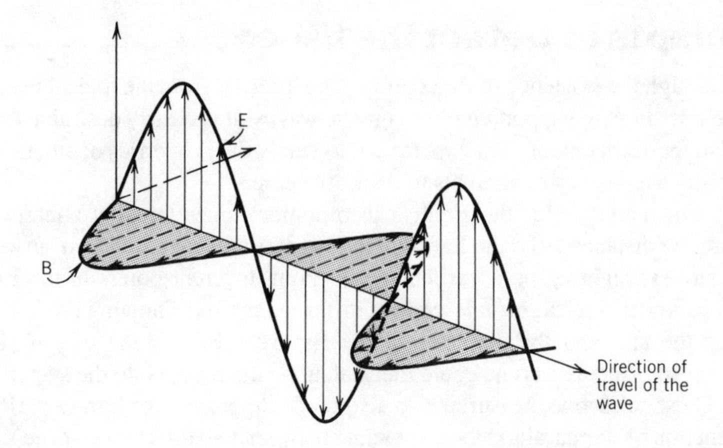

Figure 6-6. Electromagnetic wave. Variation of electric field *E* and magnetic field *B* along direction of travel for a plane electromagnetic wave.

exerted by a magnetic field on a moving charge. Additionally, it might be possible to settle once and for all the question that had plagued astronomers as to which of the heavenly objects are actually moving and which are at rest. In order to use the ether as a frame of reference, however, we must have some knowledge of its physical nature.

The ether was considered to fill all space, even empty space, because light travels across empty space. It was also considered massless. For this reason it was called the imponderable ether (ponderous means "heavy"). It was also considered to offer no resistance to motion of objects through it, because otherwise the various heavenly objects would have gradually lost speed and come to rest, or at the very least showed some similar systematic changes of their orbits with time. Because light waves are transverse waves, the ether had to have the properties of an elastic solid. The mathematical theory of elasticity then required that the ether should be very "stiff," because waves travel faster in stiff media than in soft media. Because the speed of electromagnetic waves is very high compared with other waves (by a factor of almost a million), the assumed stiffness had to be quite large. The ether had to be considered at rest with respect to absolute space, at least in the regions of space between various astronomical bodies. It was also thought that a large, massive body might "drag" some ether along with it in its motion, although this would have certain consequences for the propagation of light in the vicinity of various massive objects.

In view of all these properties, the ether clearly had to be regarded as a wondrous substance—much superior to anything created for science fiction! But, most important for the concerns of this chapter, the ether (as pointed out above and originally by Maxwell) offered the possibility of defining an absolute frame of reference, if only it could be detected.

Attempts to Detect the Ether

Although the existence of the ether may be inferred from the logical necessity to have a medium to support electromagnetic waves, it is clearly desirable to have some independent evidence or experiment to verify the properties of ether. There is a rather long history of attempts to detect the ether.

Starting about 1725, the English astronomer James Bradley attempted to measure the distance from the Earth to various stars, using a standard surveyor's technique. By sighting on a particular star from different points in the Earth's orbit, he expected to tilt his telescope at different angles. Knowing the angles of sight of the star and the diameter of the Earth's orbit, by the use of simple trigonometric calculations he could then calculate the distance to the star (Figure 6-7a). These telescope measurements also made it possible for him to verify the phenomenon of star parallax: the sight angle between two different stars depended upon the position of the Earth in its orbit (see Chapter 2). When tracked by the telescope, however, all the stars seemed to move in small elliptical orbits, regardless of the distances from the Earth. Bradley found it necessary to tilt the telescope an additional amount in the direction of the Earth's motion.

A possible explanation of the additional tilt can be seen with the help of Figures 6-7b and 6-7c. If the Earth were not moving, the starlight would travel straight down the middle of the barrel of the telescope from the middle of the opening at the top to the middle of the eyepiece at the bottom. Because the Earth is moving, however, by the time the light gets to the eyepiece, the eyepiece will have moved in the direction of the motion (to the right in the figure) and the starlight would no longer pass through the middle of the eyepiece. The solid line in Figure 6-7b shows the position of the light and telescope just as the light from the star reaches the top of the telescope; the dotted lines show the positions when the light reaches the bottom of the telescope. However, if the telescope were tilted so that the eyepiece is somewhat "behind" the opening at the top of the telescope, the middle of the eyepiece will arrive at the light beam just in time for the light to pass through, as shown in the dotted portion of Figure 6-7c. (Contrast the behavior of the beam of light with that of the object dropped from the mast of a moving ship, described early in this chapter.) The angle of tilt is proportional to the ratio of the speed of the telescope (that is, of the Earth) through the ether to the speed of light, and thus this tilt angle could be used to calculate the speed of the Earth with respect to the ether.

Some 150 years later a further check of this explanation was carried out. If the tube of the telescope were filled with water, the angle of tilt should be expected to change substantially because light travels in water at three fourths the speed of light in air. Therefore it was expected that the tilt angle should increase by a factor of 4/3 but, in fact, the tilt angle did not change at all. This experiment led to the conclusion that somehow the water "dragged" the ether along with it enough to

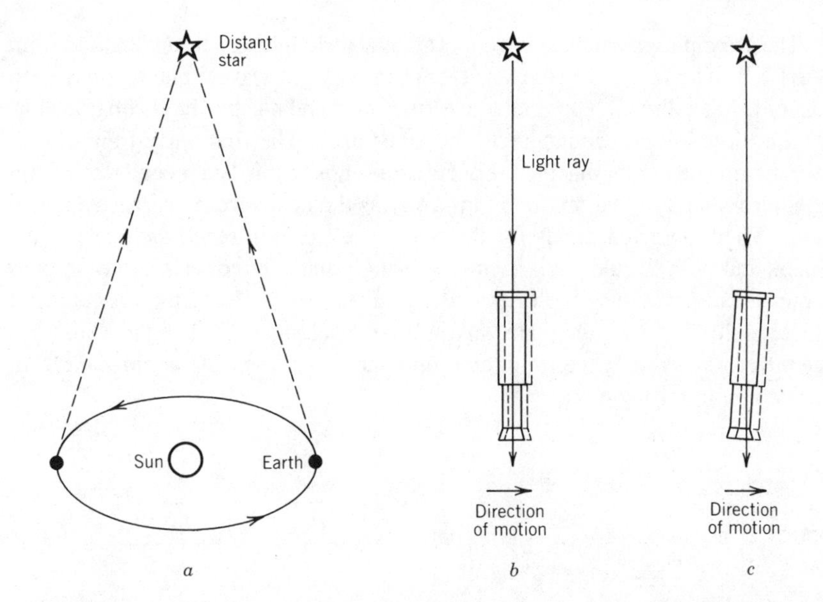

Figure 6-7. Star parallax and aberration. (a) Varying angle of view of distant star as seen from different points of Earth's orbit. (b) Motion of telescope across incoming ray of starlight. (c) Tilt of telescope to compensate for motion.

compensate for the expected effect on the tilt angle. A special experiment done some twenty years earlier implied that flowing water would indeed drag ether along with it, but not as much as needed to explain the failure of the experiment.

Starting in 1881, Albert A. Michelson, a young American, began a series of experiments intended to apply an extremely sensitive technique for measuring the motion of the Earth through the ether. For the purposes of the experiments, the important thing was to measure the relative motion of the Earth and ether; that is, one could imagine that the Earth was at rest and a stream of ether was slowly drifting past the Earth even though it was understood that it was "really" the Earth that was moving through the stationary ether. For this reason, such experiments and others like them are called ether-drift experiments.

The basic principle of such experiments can be understood by analogy with airplane flight. When an airplane has a tailwind its ground speed (its speed relative to the ground) is greater than when flying through still air; when it turns around and flies into the wind, its ground speed is less than in still air. The difference in ground speeds for downwind and upwind flight is equal to twice the speed of the wind. In Michelson's experiments, a beam of light plays the role of the airplane and the ether drift the role of the wind. Michelson also found it more convenient to compare crosswind effects with upwind/downwind effects.

The principles of Michelson's experimental technique can be understood from Figure 6-8. The idea is to compare two trains of light waves: one that has been transmitted in a direction parallel to the ether drift and one that has been transmitted in a direction perpendicular to the ether drift. The first wave-train is sent downstream to a mirror that then reflects it back upstream. The second wave-train is sent cross-stream to a mirror as far away as the downstream mirror and back again. When the actual times for the two wave-trains to reach their respective mirrors and be reflected back to their starting points are compared, taking into account the effect of the ether-drift velocity on the speed of light, the overall result is that the downstream-upstream wave-train takes longer to make the round trip than the cross-stream wave-train (assuming the two mirrors are equidistant from their common starting point).

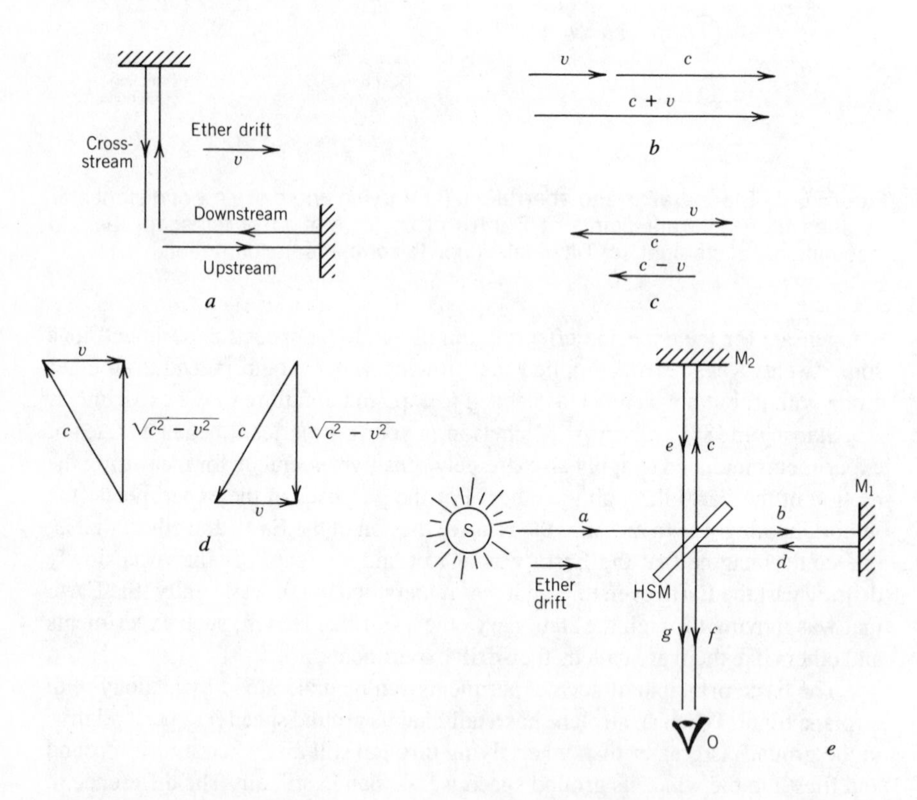

Figure 6-8. Michelson-Morley ether drift experiment. (a) Light beams traveling parallel and perpendicular to ether drift. (b) Vector addition of "downstream" velocities. (c) Vector addition of "upstream" velocities. (d) Vector addition of "cross-stream" velocities. (e) Schematic experimental arrangement. *HSM*, half-silvered mirror; *M1*, fully silvered mirror 1; *M2*, fully silvered mirror 2; *S*, light source; *O*, observer; *a, b, c, d, e, f, g,* light rays.

Michelson first performed the experiment in 1881 while he was serving as U.S. naval attaché in Germany. He repeated the experiment in 1887 in Cleveland (at which time he was a professor at Case Institute of Technology) with the collaboration of Edward W. Morley, a professor at Western Reserve University. They used apparatus twenty times as sensitive as the 1881 apparatus. Morley continued to repeat the experiment over a period of years at different positions of the Earth's orbit. The experiment is now called the Michelson–Morley experiment.

The sensitivity of the 1887 apparatus was so great that it could have detected an effect smaller than expected by a factor of 40; however, the experimental result showed no difference in the round-trip times for the two wave-trains! The experiment has since been repeated by many other investigators, with sensitivities up to twenty-five times greater than that of the 1887 apparatus but with the same result.

One obvious conclusion to be drawn from the result, called a *null result* because no difference was found between the two wave-trains, was that the Earth drags some of the ether along with it in the same way an airplane drags some air along with it in the course of its flight. As mentioned above, the measurements of star aberration with a water-filled telescope seemed to suggest that the ether was being dragged along close to the surface of the Earth.

Because the null result of the Michelson–Morley experiment implied that the ether was being dragged along by the moving Earth, it was apparent that a direct test of the ether-drag hypothesis was necessary. In 1893 a British scientist, Oliver Lodge, reported the result of an experiment designed to simulate the Earth's drag on the ether. If a very massive concrete disk is set into rotation, it should drag the ether in its immediate vicinity into rotation as well. Another less massive stationary disk placed just below the rotating disk would then have an artificially created circulating ether drift just above its surface. Lodge arranged a set of mirrors on the lower disk to produce two light wave-trains, one traveling counterclockwise around the lower disk and the other in a clockwise path. One of the wave-trains would be traveling in the same direction as the artificial ether drift, and thus its velocity relative to the mirrors would be increased, whereas the wave-train traveling in the opposite direction would have its velocity decreased. The travel time for two wave-trains could than be compared through the interference phenomenon (see Chapter 7) and the drag-induced artificial ether drift could be measured. Contrary to what was expected as a result of the Michelson–Morley experiment, a null result was again found! There was no detectible difference between the two oppositely circulating light waves.

The Michelson–Morley experiment showed that the drift of the ether (the motion of the Earth through the ether) could not be detected. The Lodge experiment showed that a direct measurement of the drag of the ether by a massive moving object also could not be detected. These two experiments apparently contradicted each other. In fact, neither experiment had succeeded in demonstrating the presence of the ether. Moreover, no other experiments have succeeded in showing ether drift or drag, or even the presence of the ether. For example, if a pair of parallel plates, one carrying positive charge and the other carrying negative charge, is suspended by a fine thread in an ether wind, there should be a resulting torque twisting the suspension. Experiment showed no such torque.

Special Theory of Relativity

In addition to the unresolved experimental problems of detecting the motion of the Earth through the ether or of even detecting the ether, some fundamental theoretical problems existed as well. Galilean–Newtonian relativity requires that forces be the same (invariant) in all inertial frames of reference. As pointed out above, however, the forces acting on electrical charges in motion would seem to be different in various inertial frames of reference. For example, if two positive charges are both traveling at the same velocity in the same direction, then, according to electromagnetic theory, there will be a resulting magnetic field that will act on the moving charges, thereby causing an attractive force between them. This force increases as the velocity increases. Yet velocity is a relative quantity, depending on the frame of reference, so the attractive force is relative and depends on the frame of reference, in contradiction to the requirement of Galilean–Newtonian relativity. Several attempts to resolve this and similar problems as well as the problems arising from the failure of the ether experiments can be looked upon as precursors to Einstein's solution to the problem.

In 1892, two physicists, G. F. Fitzgerald in Ireland and H. A. Lorentz in The Netherlands, independently suggested a solution to the dilemma posed by the Michelson–Morley experiment. They suggested that the part of the apparatus which carried the upstream-downstream wave-train (the horizontal path in Figure 6-8) became shorter by just enough to compensate for the extra time required for the light to travel along that path. Thus it would take no longer for light to travel downstream and back as cross-stream and back, thereby explaining the null result. Such an explanation can be rationalized by saying that moving through the ether generates a resisting pressure, which compresses the apparatus just as a spring is compressed by pushing it against a resisting force. The amount of contraction (shortening) necessary is very small, about half a millionth of 1 percent, so that it would normally not be noticed. The Michelson–Morley experiment, however, was sensitive enough to detect such a small effect.

Lorentz was later (1899) able to justify this postulated contraction by pointing out that it was a consequence of some ideas he had been considering in connection with the problem of the relative nature of the force between moving electric charges. He had been looking for some new transformation equations between different inertial frames of reference which would make the formula for the electromagnetic force invariant. One of the consequences of these new transformation equations, known as the Lorentz transformation, was that moving objects would contract by an amount depending upon the speed. Lorentz also found that time intervals measured at moving objects would become larger than expected. He called such expanded time intervals local time.

A French theorist, Henri Poincaré, even suggested in 1904 that it was futile to attempt to measure the motion of the Earth—or anything else—with respect to the ether. Using the idea that it was impossible to measure velocity in an absolute sense, he went beyond Lorentz and showed that the mass of an object (that is, inertia, the aspect of the mass that plays a role in Newton's second law of motion) increases as the object's speed increases. He also showed that there is a maximum speed any object can attain: the speed of light.

All of these ideas are included in Einstein's special theory of relativity. The significance of Einstein's work was that he was able to show simply and directly (at least from a physicist's viewpoint) that they were natural consequences of a profound and insightful reexamination of some basic assumptions about the nature of physical measurements, whereas Fitzgerald, Lorentz, and Poincaré introduced them in order to deal with a specific problem. Einstein's thinking was very much influenced by the Austrian physicist-philosopher Ernst Mach, who had carried out a critical and incisive reexamination of Issac Newton's *Principia*. Mach was one of the founders of the logical positivist school of philosophy and was particularly critical of Newton's definitions of absolute space and time and of mass. (Mach was not the first person to criticize Newton's work, of course. The German philosopher Leibnitz, contemporary and rival of Newton, had been rather critical of these same definitions.)

The circumstances surrounding the development of physics at the beginning of the twentieth century were somewhat similar to those at the time of Isaac Newton. Several very talented scientists were close to making a breakthrough on a class of problems that were ripe for solution, but only one of them—an Einstein or a Newton—was able to master the situation. As an example, the Lorentz-Fitzgerald contraction ostensibly accounts for the failure of the Michelson-Morley experiment; in fact, there is a modified version of that experiment (called the Kennedy-Thorndyke experiment) which was designed to circumvent the Lorentz-Fitzgerald contraction and also gives a null result. Moreover, as we will see below, Einstein's approach provides an entirely different interpretation of the failure of all the ether experiments.

Probably no other scientist, except perhaps Isaac Newton, has been as famous as Albert Einstein (1879-1955). Even three decades after his death he is as renowned for his white hair, droopy mustache, curved pipe, stooped posture, and gentle informality as for his scientific theories. (Figure 6-9 shows Einstein as a young man, when he first began to make his mark on the scientific world.) As is often true of well-known people, there are a number of myths and misconceptions about him and his work. Although he began speaking later than usual as a young child, his considerable abilities were quite evident even in childhood. He did not do as well in school subjects of no interest to him as in those he enjoyed; but when

Figure 6-9. Einstein in 1905 (Reproduced by permission of Lotte Jacobi, Hillsboro, N.H.)

he found that he had to master some material in order to meet entrance requirements to the school he wanted to attend, he promptly did so. He particularly disliked regimentation but was capable of considerable self-discipline. He was not an outstanding mathematician but was competent in the mathematics he needed.

He made significant contributions in many areas of physics besides relativity theory, publishing more than 350 papers in his lifetime. Not all of his work was in theoretical physics, and he even filed some patent claims. Unlike Julius Robert Mayer, who had great difficulty communicating his important ideas on energy because he did not frame his ideas in terms his contemporaries could follow, Einstein was quite successful in communicating his most controversial ideas. He readily participated in the scientific debates of his day with good humor.

He did not hesitate to speak out on political matters that he considered important. He also deplored the use of the physical theory of relativity to justify moral relativity, if only because such use betrayed a misunderstanding of the true meaning of the theory.

So far as the theory of relativity is concerned, some people have questioned the uniqueness of Einstein's contributions. He himself conceded that, considering the scientific ferment of the time, Poincaré might well have developed the special theory if Einstein had not. He stated, however, that he began thinking about problems related to relativity theory when he was sixteen years old. Moreover, Einstein's approach to the subject was characteristically straightforward and elegant, cutting through to the essential ideas. The general theory of relativity, discussed later in this chapter, is universally recognized as Einstein's unique creation.

In 1905, while working as an examiner in the Swiss patent office, Einstein published three very important papers, one of which was entitled "On the Electrodynamics of Moving Bodies" (the other two papers are mentioned in Chapter 4 and Chapter 7). In this paper he was not so much concerned with relative quantities, whose magnitudes depended on the frame of reference from which they were measured, but rather with invariant quantities, which would be the same in all inertial frames of reference. He enumerated two special principles that should be applicable in all frames of reference:

I. The laws of physics are invariant in all inertial reference frames.

This means that the formulas expressing the various laws of physics must be calculated in the same way in all inertial reference frames.

II. It is a law of physics that the speed of light in empty space is the same in all inertial reference frames, independent of the speed of the source or detector of light.

Einstein pointed out that this meant that the ether could not be detected by any experimental means, and therefore it was a useless concept which should be discarded.

The first principle states simply that a relativity principle does exist, as Galileo, Newton, and others had originally indicated. It particularly emphasizes that the principles of physics are the same everywhere. The second principle is the important new physical insight.

The constancy of the speed of light is a significant result of all the experiments aimed at detecting the ether. Equally significant, these experiments confirm Galileo's idea that it is not possible to determine absolute motion. This also means that measurements in one frame of reference are just as valid as measurements in any other frame of reference. Questions such as "How fast is the Earth really moving?" are therefore meaningless.

Under the influence of Mach, Einstein realized that it was necessary to reconsider the meaning of space and time and how they are measured. He recognized that space and time are not independent concepts, but are necessarily linked with each other. Moreover, these concepts are defined by measurement. For example, time is measured by watching the movement of the hands of a clock or the passage of heavenly bodies through the sky. We know time has passed because we see that these objects have changed their positions in space. The speed of light is involved because these observations are made by virtue of the fact that light travels from the moving objects to our eyes. (It does not matter if electrical signals are used to tell us that the objects move: such signals also travel with the speed of light, as shown from electromagnetic theory.)

Einstein showed that, in a manner of speaking, time and space are interchangeable, as is illustrated by the following set of statements, which exhibit the symmetry of space and time:

> *I. A stationary observer of a moving system will observe that events occurring at the* ***same place*** at ***different times*** in the moving system occur at ***different places*** in the stationary system.

> *II. A stationary observer of a moving system will observe that events occurring at the* ***same time*** at ***different places*** in the moving system occur at ***different times*** in the stationary system.

> *III. A stationary observer of a moving system will observe that events occurring at the* ***same time*** at the ***same place*** in the moving system occur at the ***same time*** and ***same place*** in the stationary system.

Statement II is obtained from Statement I by interchanging the words *time* and *place*. This interchange changes the meaning of the statement. On the other hand, interchanging the words *time* and *place* in Statement III does not change the meaning of that statement.

To illustrate these statements, the moving system might be an airplane traveling from New York to Los Angeles and the stationary system might be the control tower of an airport on the Earth. An airline passenger might be sitting in seat 10C. At 8:00 A.M. the passenger is served orange juice, while the airplane is above Albany, New York, and at nine o'clock the passenger is drinking a cup of coffee after breakfast, while the airplane is passing over Chicago. In the moving system, the airplane, both events occurred at the same place, seat 10C, but at different times. In the stationary system, the Earth, the two events occurred at different places, over Albany and over Chicago, as would be seen if an observer in the control tower could look inside the airplane.

The foregoing scenario is very plausible, but a scenario based on the second statement is implausible: Sometime later, when the airplane is over Denver, Colorado, the passenger, who is reading a physics book, looks up and sees a federal marshal at the front of the airplane and a hijacker at the back of the airplane, with guns pointed at each other. Both guns are fired at the same time, as seen by the passenger. As seen by the observer in the control tower on the Earth, however, the shots were fired not simultaneously but at different times. Implausible as it seems, the scenario based on the second statement is correct.*

In a third scenario, after both shots miss, the passenger notices that the flight attendant standing next to him simultaneously gasped and dropped a pot of coffee in his lap, in seat 10C. The president of the airline, watching from the control tower, sees that indeed the flight attendant simultaneously gasped and dropped the pot of coffee into the passenger's lap, over Denver.

The point of all this is to illustrate that because space and time are intertwined they are relative quantities that are different in different inertial frames of reference. Events that are simultaneous in time in one frame of reference may not be simultaneous in time in another inertial reference frame. Only if the simultaneous events occur at the same place, as in the third scenario, are they simultaneous in all inertial frames of reference.

Einstein showed that time intervals between two events will be measured differently in different inertial reference frames. Even two otherwise identical clocks will run at different rates in the two reference frames; that is, the time between tick and tock will be different. This can be illustrated by a rather simple example called a mirror clock, shown in Figure 6-10a, which shows a light bulb inside a box having a coverable hole and a mirror placed 93,000 miles above the hole. If the hole is momentarily uncovered and then recovered, a brief flash of light will arrive at the mirror a half-second later and then be reflected back to the covered hole, arriving yet another half-second later. Two events have taken place: (1) the emission of the light flash from the hole and (2) the return of the light flash to the covered hole one second later. This mirror clock is then a one-second timing clock.

If two such clocks are made, one could be placed in inertial reference frame A and the other placed in inertial reference frame B. The two reference frames are moving at constant velocity relative to each other. As seen in its own reference frame, each clock works perfectly (the speed of light is the same in all reference frames). As seen from the other reference frame, however, each clock runs slow, as shown in Figure 6-10b. The clock in reference frame A, as seen from B, moves with a speed of v. During the half-second it takes the light to travel from the hole in the box to the mirror, the mirror has moved from its original position at the left of the figure to the dotted position in the center of the figure. As seen from B, therefore, the ray of light traveled upward along a diagonal path until it was reflected, then downward along a diagonal path. (Note the similarity between this

*The time interval between the two shots is very, very small, certainly much less than human reaction times, on an airplane flying at normal speeds. But if the airplane were flying at a very high speed, say half the speed of light, the time interval between the two shots as observed from the control tower would be quite noticeable.

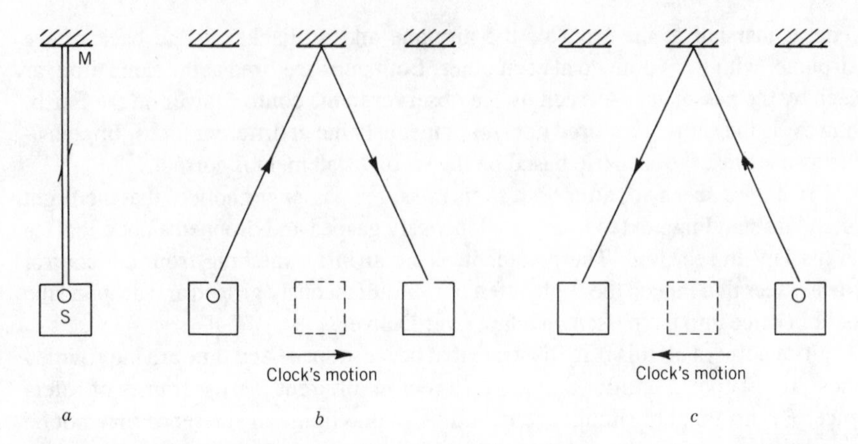

Figure 6-10. Light rays in mirror clock. *S*, light source, *M*, mirror. (a) As seen from reference frame in which clock is at rest. (b) Clock in reference frame A, as seen from reference frame B. (c) Clock in reference frame B, as seen from reference frame A.

situation and that of the object dropped from the mast of a moving ship, discussed earlier.) To an observer in reference frame B, the flash of light in the clock of reference frame A has traveled a greater distance than the flash of light in his own stationary clock. Because the speed of light is constant for all frames of reference (Einstein's second special principle), then the time interval between the two events—the emission of the light flash and its return—is not one second for the moving clock but some longer time. The moving clock runs slow!

The critical element in this analysis is the assumption that the speed of light is constant in both frames of reference. This is the important new fact indicated by Einstein's special theory of relativity and eventually leads to some particular predictions that can be tested experimentally.

The time interval measured by the moving clock is calculated by dividing one second by $\sqrt{1 - (v/c)^2}$, where v is the relative velocity of the reference frames and c is the velocity of light. (This formula can be justified from a simple geometric calculation.) If, for example, the ratio of v/c is 3/5, the time interval on the moving clock as observed from B is 1.25 seconds. The moving clock loses 1/4 second every second.

An observer in reference frame A, however, sees the situation quite differently. Relative to A, it is the clock in B that is moving (Figure 6-10c), and the clock in B runs slow. There is no contradiction here—the observer in reference frame A "sees" time intervals between events differently than the observer in B. Because all motion is relative, each observer may say that his or her reference frame is at rest and the other reference frame is moving with respect to him or her. Both observers can validly claim that the clock in the other reference frame is running slow. This apparent paradox of the equally valid conflicting claims will be discussed further and clarified below.

The phenomenon exemplified by the slow-running clock is called *time dilation*. There is another effect observed in moving inertial reference frames, called *length contraction*, that is similar to the Lorentz–Fitzgerald contraction. If a measuring stick is placed in reference frame A and another identical measuring stick placed in reference frame B, both sticks will be unaffected by the motion as seen by observers in the same reference frame with each stick. However, if the sticks are placed parallel to each other and to the direction of relative motion of the two reference frames, when the observer in A measures the stick in B, in comparison with his or her own stick the one in B will appear to be shorter. Similarly, when the observer in B measures the stick in A, the stick in A will appear shorter. Which stick is really shorter? They both are! (An alternative answer is neither one. The question is really meaningless, as we will find in the next few paragraphs.)

Both observers will find that the other stick (the one in relative motion with respect to the reference frame) has undergone length contraction. The stick that is at rest with respect to the reference frame maintains its proper length. If the sticks are placed perpendicular to the direction of motion of the reference frames, there will be again no change in their length (they will be narrower, however, as observed from the other reference frame). Any object in motion in a given reference frame will undergo contraction in that dimension which is parallel to the direction of the motion.

The interrelationship between time dilation and length contraction is shown most clearly by introducing time as a "fourth dimension"; rather than considering space and time as distinct entities, they should be regarded as different aspects of one unified entity called space-time. Instead of discussing a particular location in space or a given point in time, it is more useful to consider events in space-time. An event is specified by describing both where and when it happened or is going to happen. In space alone, the distance between two locations is measured. In time alone, the elapsed time between two occurrences is measured. In space-time, however, the space-time interval between two events is measured.

One way to understand this is to consider the analogous relationships between a stick and its shadows (see Figure 6- 11). If a stick is held at some angle with the horizontal above a table and next to a wall, it will cast shadows on the table and on the wall when properly illuminated. If the stick is illuminated from above, a horizontal shadow will be cast on the table; if it is illuminated from the side, a vertical shadow will be cast on the wall. The length of the shadows on the table and on the wall will depend upon the angle at which the stick is held. If the stick is almost vertical, the table shadow will be short and the wall shadow long; whereas

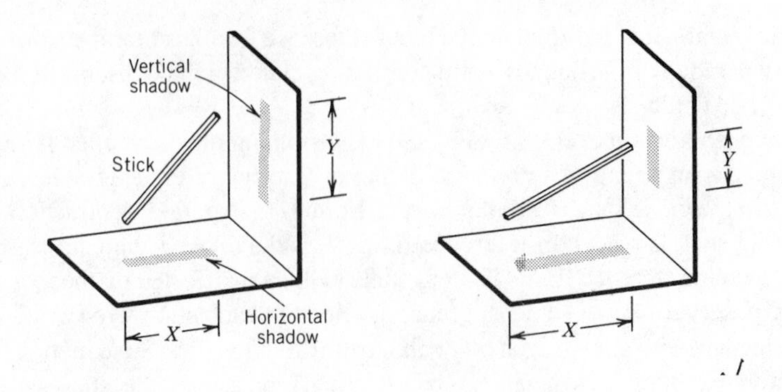

Figure 6-11. Horizontal and vertical shadows, X and Y, of a stick. Regardless of angle of stick, its length can be calculated from $\sqrt{X^2 + Y^2}$

if the stick is almost horizontal, the table shadow will be long and the wall shadow short. The length of the stick will be unchanged (invariant); only its shadows will change as its angle is varied.

If it were impossible to measure the length of the stick directly, it would still be possible to calculate its actual length from the lengths of the two shadows. If the length of the table shadow is X and the length of the wall shadow is Y, then according to the Pythagorean theorem the true length of the stick is $\sqrt{X^2 + Y^2}$ No matter how the stick is angled with resulting changes in X and Y, the length of the stick will always be the same, because the length of the stick is invariant, no matter what the angle.

In terms of space-time, when measurements are made of a distance between two events and an elapsed time between those same two events, the quantities being measured are the "space-shadow" and the "time-shadow" of the space-time interval between the two events. Einstein showed that if T is the elapsed time, L is the distance between the two events, and c is the speed of light, the space-time interval between the two events is equal to $\sqrt{c^2 T^2 - L^2}$. Different observers in different frames of reference may measure T and L to have different values (this corresponds to changing the angle of the stick), but when they calculate the space-time interval using their own measurements, they will all get the same numerical result. The space-time interval is an invariant and does not change from one inertial frame of reference to another inertial frame of reference.

The space-time interval cannot be measured directly, but it can be calculated from the measurements. The "space-shadow" and the "time shadow" are the quantities that are measured, but the values of the measurements are relative. It is as meaningless to ask what is the correct time interval or the correct distance between two events as it is to ask what is the "real length" of the shadow of an object.

These new concepts of space and time introduced by Einstein have a number of consequences. From these concepts he was able to show that it is necessary to use the new transformation equations that had been suggested by Lorentz. These equations, the Lorentz transformation, lead to new formulas for calculating relative velocities of objects in different frames of reference. Early in this chapter an example was given of a ball thrown forward at 60 mph relative to a ship traveling at 30 mph relative to the ocean. The speed of the ball relative to the ocean is then 90 mph. In algebraic terms, if the velocity of the ball relative to the ship is u and the velocity of the ship relative to the ocean is v, the speed of the ball relative to the ocean is $u + v$, according to the Galilean transformation, which is the name for the transformation equations used by Galileo and Newton and their successors. According to the Lorentz transformation, however, the speed of the ball relative to the ocean is given by $(u + v)/(1 + uv/c^2)$, with c being the speed of light. Carrying out the calculation according to the new formula, called the Lorentz velocity addition formula, the speed of the ball relative to the ocean will be less than 90 mph, but only by about 0.4 of a millionth of a millionth of a percent. The difference between the two calculations is immeasurably small in this case.

However, if the speed of the ball relative to the ship is 0.6 of the speed of light and of the ship relative to the ocean is 0.3 of the speed of light, then there is a substantial difference in the speeds calculated by the two different formulas. For the Galilean transformation, the speed of the ball relative to the ocean is $0.6 + 0.3 = 0.9$ of the speed of light. In the Lorentz transformation, the speed of the ball relative to the ocean is 0.76 of the speed of light. If the component speeds are 0.9 and 0.6 of the speed of light, the Galilean formula gives $1.5c$, whereas the Lorentz formula gives $0.97c$. In fact, as Table 6-2 illustrates, it is impossible to add two speeds that are even only slightly less than the speed of light and obtain a result that will be equal to the speed of light. If one of the speeds is exactly equal to the speed of light, then adding it to the other speed according to the Lorentz velocity addition formula will yield the speed of light. This means that the speed of light is constant for all observers, as demanded by Einstein's relativity principles.

Thus the speed of light turns out to be a natural speed limit because it is impossible, according to the Lorentz velocity addition formula, to add some additional increment of speed to an object to get it over the speed of light. This means that Newton's laws of motion as usually discussed, in particular Newton's second law, are incorrect. As discussed in Chapter 3, if a constant force is applied to an object, the acceleration of the object is constant; that is, a graph of velocity versus time gives a straight line increasing continually with time. However, the imposition of a maximum speed limit means that the velocity cannot increase indefinitely. The velocity may become very close to the speed of light but may not exceed it, as shown in Figure 6-12. As the velocity gets closer to the speed of light, it increases at a smaller rate; that is, its acceleration becomes smaller. The object behaves as if its inertia is increasing because the same force is giving less

Table 6-2 Comparison of Results for Velocity Addition According to Galilean and Lorentz Transformations

u	v	u + v	$\frac{u + v}{1 + uv/c^2}$
60 mph	20 mph	90 mph	90 mph
186 mps (0.001c)	18.6 mps (0.0001c)	204.6 mps	204.59998 mps
0.6c	0.3c	0.9c	0.763c
0.5c	0.5c	c	0.800c
0.75c	0.75c	1.5c	0.960c
0.9c	0.6c	1.5c	0.974c
c	0.1c	1.1c	1.000c
c	c	2c	c

acceleration. But because mass is the measure of inertia, it can be said that the mass of the object becomes greater as its speed increases. The mass increase becomes very noticeable as the speed comes very close to the speed of light.

Velocity is a relative quantity, however, so that if the object were observed from an inertial frame of reference traveling at the same speed as the object, then its velocity would be zero and in that frame of reference its mass will be unchanged. Thus mass is also a relative quantity.

Actually, Newton's laws of motion are not irreparably invalidated by Einstein's relativity theory. Rather, the definitions of some of the quantities that are calculated in mechanics must be changed. It is necessary to restate Newton's second law in its original form, making force proportional to the time rate of change of momentum, and specifically recognize that in calculating the momentum as the product of mass and velocity the mass is not independent of velocity. Similarly, it is necessary to reconsider and modify the definition of kinetic energy to include the fact that increased kinetic energy does not come from increased velocity alone but from increased mass as well. The work expended to give an object kinetic energy not only increases the speed of the object but also its mass. (The new formula is not just $1/2\,mv^2$ with m also increasing, but is somewhat more complicated.) In other words, mass is a form of energy, just as heat is a form of energy. Where Poincaré and others had shown that electromagnetic energy can be regarded as having mass, Einstein boldly asserted by his formula $E = mc^2$ that all mass is equivalent to energy. He pointed out that not only did the relativistic increase in mass of an object as its speed became closer to the speed of light

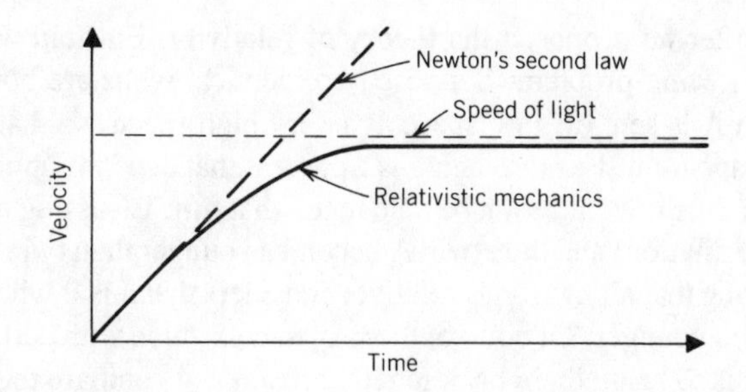

Figure 6-12. Graph of velocity versus time for constant force. In relativistic mechanics, the velocity cannot increase indefinitely, but rather is limited by the speed of light.

represent an increase of energy, but even the mass of an object at zero velocity, called its proper mass, or rest mass, is a form of energy.

In principle, according to Chapters 4 and 5, energy can be converted from one form to another. Thus mass energy can be converted into light energy or gamma-ray energy or vice versa. Such conversions are involved in nuclear energy and in the creation and annihilation of elementary particles of matter, as will be discussed in Chapter 8.

One important point needs to be emphasized regarding the implications of the theory of relativity for other theories of physics. Relativistic effects usually depend upon extremely high velocities to be observable. Typically a speed equal to 10 percent of the speed of light will result in a one-half-percent effect. The effect of a speed that is 1 percent of the speed of light will be a hundred times smaller, and not measurable in most cases. Even such a speed is extremely high, 1860 miles per second, compared with the speed of supersonic rockets, which conceivably could be 1860 miles per hour—that is, slower by a factor of 3600. As a result, in most common cases, relativistic effects are not noticeable. On the other hand, relativity plays a large role in astrophysics and in atomic and nuclear physics.

It is interesting to speculate how things might appear to us if the speed of light were much less than it actually is—say 20 mph. The consequences would be interesting: For example, traffic citations for speeding would be almost nonexistent because it would be impossible to travel faster than 20 mph. Such a situation is the subject of a series of short stories written mostly in the 1930s and 1940s by G. W. Gamow.* In these stories, bicycles and their riders become extremely short and thin, respectively, as they pedal along, as seen by pedestrians, whereas city blocks become extremely short as seen by the bicycle riders.

*Some aspects of these stories distort relativity theory. In particular, three-dimensional objects moving at speeds close to the speed of light appear rotated rather than foreshortened in the direction of motion. Nevertheless, the general thrust of the stories is useful in visualizing the effects of relativity, as well as those of quantum theory, the subject of Chapter 7.

Sometime after he proposed the theory of relativity, Einstein was asked to consider the following problem: Suppose two identical twins are born, and at a certain age twin A is sent off into space at a very high speed, so that relativistic effects become apparent. Because aging is a process that depends upon time, twin B, who stays on Earth, would soon be able to tell that time is passing more slowly for twin A (time dilation) and thus twin A becomes younger than twin B. Twin A, however, knowing that all motion is relative, considers that it is B who is moving, and therefore B is younger. Of course, the way to determine with certainty which twin is younger is to bring them back together again and compare them with one another. A's spaceship reverses course and returns to Earth so that the comparison can be made. Einstein was asked which twin was younger and why. After some thought, Einstein replied that the traveling twin, A, would indeed be younger. (An experiment to test this has actually been carried out. Carefully stabilized atomic clocks have been sent on long airplane trips and then returned for comparison with identical clocks that were kept in a laboratory. The traveling clocks turned out to be younger in that they had counted off less time intervals [because of time dilation] than the fixed clocks.)

The crux of the distinction between the ages of the twins is that twin A was subject to accelerations (1) when he was sent off into space and (2) when he reversed direction and returned to Earth. In Einstein's general theory of relativity (to be discussed below in terms of all reference frames and not just inertial reference frames) it is shown that accelerations result in a slowing of time. Without going into further detail, we can see that there is a difference in the history of twin A as compared with that of twin B. This is nicely illustrated with the aid of a Minkowski diagram.

The Minkowski diagram is simply a graph of time versus distance, as shown in Figure 6-13a. Note that in this graph, unlike those shown earlier in Chapter 3, the X axis is plotted horizontally, and the vertical axis is the cT axis rather than the T axis. (The cT axis is used to emphasize the unity of space-time and to give the so-called fourth dimension the same kind of measurement units as the other three dimensions.) The data plotted on such a diagram are the positions of an object and the particular times it has those positions. An object at rest in a frame of reference will keep the same value of X as time goes on, and the resulting graph will be a vertical straight line, as indicated by the line B in the diagram. An object traveling to the right is represented by the lower part of line A in the diagram, and one traveling toward the left by the upper part of line A.

Such a graph is called a world line, and in effect tells the history of the object because it records where the object has been and when it has been there. Note that an object traveling at the speed of light would be shown by the dotted line C, which is at a 45-degree angle. Thus it is impossible, according to Einstein's relativity principles, for a world line to have a slope less than 45 degrees. If A and B are the

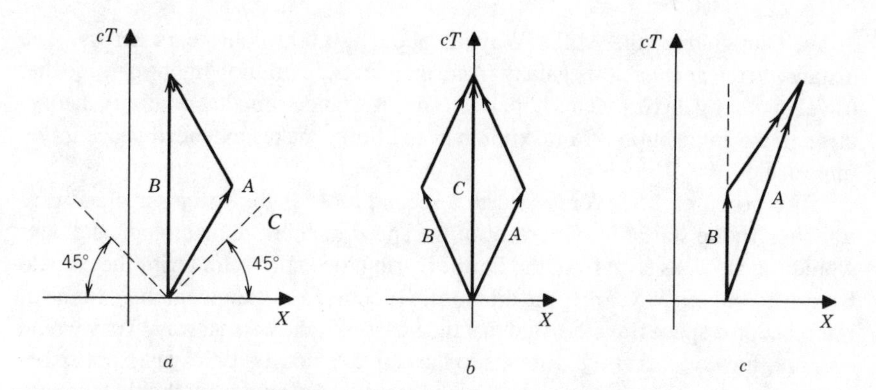

Figure 6-13. Minkowski diagrams. *cT*, time axis; *X*, space axis. (a) A leaves Earth and returns later. (b) Both A and B leave Earth and return later. (c) A leaves Earth, followed by B sometime later, who "catches up" with A.

world lines of the two identical twins, then Figure 6-13a shows that their histories are different, so we can predict that their ages will be different when they are brought together again.

Figure 6-13b portrays a somewhat different situation. Instead of identical twins, it shows the case of identical triplets A, B, and C. While C stays on Earth, A travels to the right as before and B travels to the left with respect to the Earth. When both A and B return to Earth, their ages will be identical, but less than that of C. Yet another situation is shown in Figure 6-13c, which deals only with twins. In this case, instead of returning A to Earth, B is sent off on a spaceship to catch up with A, and this time B turns out to be younger when they meet.

Relativity and space travel have greatly influenced the popular imagination. Novels, motion pictures, and television shows often depict space travel, starting with Jules Verne well over a century ago. Similarly, the idea of time as a fourth dimension, and of travel through time, goes back at least as far as Mark Twain, even before Einstein appeared on the scene. Two questions naturally arise: Is it possible to travel to distant stars or galaxies? Can one travel in time?

Astronomical distances are measured in light-years; that is, the distance that light would travel in one year. In order to travel to a star that is 100 light-years away, a trip of at least 100 years is required even for a spaceship traveling at the maximum speed, the speed of light. If the crew of the ship sent back a radio message immediately upon arriving, it would take another 100 years for the mission controllers on Earth to receive the message, so both the crew and the mission controllers would have to be extremely long-lived. The distance to the nearest star (Alpha Centauri) is 4.3 light-years, so the minimum time for a round trip would be 8.6 light-years. Most stars, however, are more than 100 light-years away. Our galaxy, the Milky Way, is about 120,000 light-years across. The distance to the nearest other galaxy (Andromeda) is 2.2 million light-years, so that travel to other galaxies seems hopeless. Thus it would seem that relativity theory, through the imposition of a maximum speed limit, makes extended space travel impossible.

The problem, however, is worth a second look. If the space travelers were traveling to the star at 99.9 percent of the speed of light, their biological clocks would run slow, as seen from the Earth. During the entire round trip, they would have aged only 9.94 years (time dilation). Of course, as seen from their frame of reference, the space travelers find that time passes in the normal way. They would observe, however, that the distance to the star was not 100 light-years but rather only 4.47 light-years, so that at their speed of travel the round trip should take only 9.94 years. They would find, of course, that things might well have changed considerably when they returned to Earth. For example, the Earth calendar would have advanced by 200 years and there might be no record of their ever having departed other than an obscure footnote in a textbook.

Thus, although space travel is in some sense feasible, time travel is another story. There is no question that people travel forward in time, but traveling backward in time does not seem to make sense, particularly if time travelers could participate in events of an earlier time. Cause-and-effect relationships would be destroyed. For example, children might be living before their grandparents ever met. Of course, when observations are made of occurrences in deep space, say in a star 100 light-years away, what is observed is something that happened 100 years ago. The crew of the spaceship discussed above cannot affect galactic events that are observed on Earth, because they have already happened. Moreover, they cannot affect events happening now or even five or ten years from now, because in terms of observations from Earth it will take them a little more than a hundred years to arrive at the star.

Referring back to the Minkowski diagram of Figure 6-13a, only those events in the part of the diagram above the two 45-degree dotted lines are accessible to travelers starting out from a given event in space-time. All other events, especially those in the negative cT direction that involve going backward in time, are not accessible.

General Theory of Relativity

The theory of relativity outlined and discussed above is known as the "special" theory of relativity, because it applies to only one special case. It gives the correct transformation equations between nonaccelerated (inertial) reference frames. With these transformation equations, the laws of physics appear in the same form in all such inertial reference frames and are seen to satisfy a relativity principle as originally expected by Galileo, Newton, and others. Using the results of these new transformation equations, we have seen that certain unexpected consequences follow in comparing times, lengths, and masses in different inertial reference frames. These transformation equations apply only between inertial reference frames, however, and Einstein believed that one should be able to use any reference frame, including accelerated reference frames. A theory that could specify the transformation equations using accelerated reference frames would not be limited to the special case of inertial reference frames; this has become known as a "general" theory of relativity.

Einstein spent the ten years or so after publishing his special theory of relativity working primarily on a general theory of relativity. In 1916 he published his completed description of such a theory to considerable reaction from the scientific community. Although it is said that any one of several other physicists were about to arrive at the special theory of relativity when Einstein published his work in 1905, it is generally accepted that Einstein's work on general relativity was far ahead of any of his contemporaries. Einstein's general theory of relativity stands clearly as one of the greatest achievements of the human intellect.

Einstein began his work on the general theory of relativity with an observation which, at first, seems like an aside. He noted that the same quantity—mass—is involved in both Newton's second law of motion and in the universal law of gravitation. He observed that if these two laws actually were independent, they would define two different kinds of mass. Newton's second law ($F = ma$) would define an inertial mass, and the law of gravity ($F = Gm_1m_2/r^2$) would define a gravitational mass. In fact, a number of precision experiments were carried out at about that time to see whether there was any difference between the inertial and gravitational masses of an object. No difference between the two masses could be measured, and it seemed a remarkable coincidence that they should be the same. Einstein questioned whether it is coincidental that the same quantity is involved in both these laws. He decided that acceleration (as involved in Newton's second law) and gravity must be related. He then proceeded to show that it is impossible, in fact, to tell the difference between a gravitational force and an "equivalent" acceleration—a relationship that has become known as Einstein's equivalence postulate.

As a simple example of the equivalence of gravity and acceleration, consider the hypothetical situation of a person standing in a spaceship in outer space, far away from any star or planet. If the spaceship is moving at a constant velocity, the person would experience weightlessness, just as the astronauts did on their trips to the Moon. If the spaceship were resting on a planet, the person would experience a force of gravity, giving him or her weight. Einstein's postulate indicates that the same effect could be produced by simply accelerating the spaceship. We are all well aware of the basic idea involved: When an elevator car starts moving upward, we feel the acceleration and are pushed down toward the floor. (This is Newton's third law of motion. The bottom of the elevator pushes up on the passenger and the passenger pushes down on the elevator.) If one were to make the acceleration of a spaceship in outer space continuous and exactly equal to the acceleration of gravity of the Earth's surface, it would create a "downward" force that would feel like gravity. If the spaceship had no windows, it would be impossible for the person inside, either by "feel" or even by careful experiment, to tell whether the force was due to an acceleration or to a gravitational force from a large mass.

Thus we have a situation where a force can be simulated by an acceleration. Many other examples exist. The acceleration does not have to be in a straight line. When a fast-moving car goes around a corner, the riders all feel a "force" pushing them toward the side of the car. This force is referred to as a centrifugal force—not a force at all but rather just the effect we feel as our inertial mass tries to continue in straight-line, uniform motion while the car goes around the corner. If there is no restraining force (so-called *centripetal* force, discussed in Chapter 3), the riders will accelerate relative to the car. Relative to the ground, however, they continue in straight-line uniform motion.

The centrifugal force is a fictitious force in that it is not caused by some external force exerted on a body but rather is a result of the acceleration of the car and the inertia of the object. Such fictitious forces are often called *inertial forces*. Note that the centrifugal force can also be used to simulate gravity. One suggested way to provide a "gravitylike" environment in space is to make a space station in the doughnut-shape of a spinning torus. Such a space station was portrayed in the popular science-fiction movie *2001*, for example. The centrifugal force resulting from the spinning torus would push everything (and everyone) toward the outer edge of the torus. If the torus were to spin at just the right rate, it could provide a centrifugal force just equal to the force of gravity on the Earth's surface.

The important thing about a centrifugal force for the present discussion is that it is actually a fictitious force, which arises only when one tries to use an accelerated reference frame. In an inertial (nonaccelerated) reference frame, this force does not appear. Because it is mass that determines the magnitude of this fictitious force (inertial mass), Einstein wondered if it could be true that gravity, which is also due to mass (gravitational mass), could also be regarded as a fictitious force the existence of which depended upon the choice of reference frame. Basically, this idea arises from his equivalence postulate—that inertial and gravitational masses are the same. Thus Einstein felt it was not coincidental that mass is the primary thing responsible for both inertia and gravity, that they must be related.

In fact, it is possible, by suitable choice of reference frame, to "get rid of gravity" even on the surface of the Earth. Suppose a person stands on a scale in an elevator that is moving with constant velocity. The scale will indicate his or her weight. If the cable attached to the elevator breaks, the elevator falls with the acceleration of gravity. While the car is falling, the scale will actually read zero, indicating that the person is weightless. In the frame of the falling elevator, the force of gravity has vanished. Similarly, astronauts are weightless when in orbit around the Earth, because their reference frame, the space capsule, is accelerating toward the Earth (they are constantly "falling").

The equivalence principle does much more than assert the identity of gravitational and inertial mass. It states that any effect which can be ascribed to an accelerated reference frame could equally well be called a gravitational effect. Figure 6-14 shows an accelerating spaceship in outer space. A ball is thrown across the ship and takes two seconds to hit the other wall. Figure 6-14b shows the location of the ball and the spaceship after one second and Figure 6-14c after two seconds, as seen from outside the spaceship. The ball has traveled in a straight line. Figure 6-14d, e, and f show the location of the ball as seen from inside the spaceship. As seen from inside the spaceship, the ball has traversed a parabolic path. Observers inside the spaceship can describe the trajectory of the ball as resulting from the combination of horizontal motion and falling motion due to gravity, as Galileo did in his analysis of projectile motion (Chapter 3 and Figure 3-5).

If the horizontal speed of the ball is increased, it will still travel a straight line as seen from outside the spaceship and a parabola as seen from inside the spaceship, although it will not "fall" as far. Even if the speed of the ball were equal to the speed of light, as seen from inside the spaceship it would still travel a parabolic trajectory and seem to fall a little. Figure 6-14g shows the effect of increasing the horizontal speed of the ball as seen from inside the spaceship.

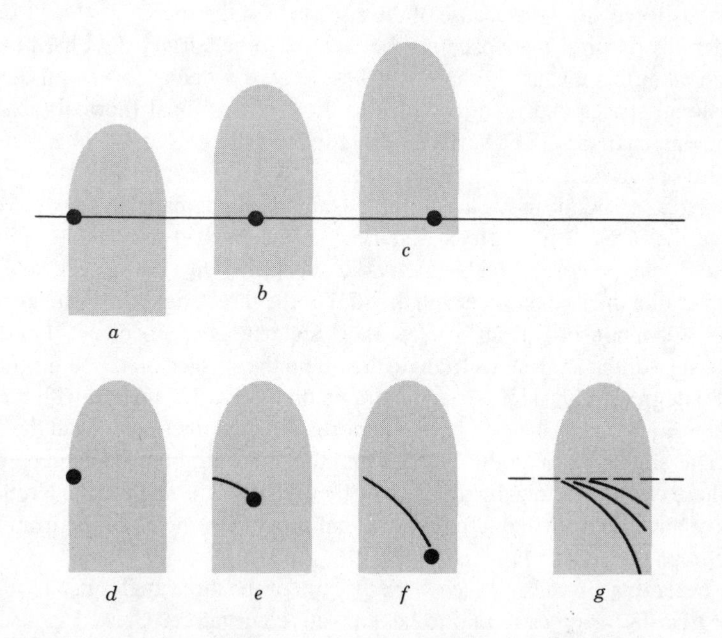

Figure 6-14. Projectile trajectory in accelerating spaceship. (a, b, c) As seen from outside the spaceship. (d, e, f) As seen from inside the spaceship. (g) Internally viewed trajectories for different horizontal velocities.

According to the equivalence principle, all of the effects shown in Figure 6-14, as seen from inside the spaceship, can be ascribed either to the acceleration of the spaceship or the effect of a gravitational field. If a ray of light were to travel across the spaceship, it too would follow the same path as the ball traveling at the speed of light. Thus, according to the equivalence principle, a ray of light can be "bent" by a gravitational field.

Einstein interpreted the bend of a ray of light as representing a curvature of space itself. Just as in special relativity, he reasoned that we only know about space through measurements with light. The path of a beam of light (really any electromagnetic radiation) reveals the nature of space. A carpenter or machinist determines whether a surface is flat or curved by measuring it with a straightedge or ruler. But how does one know whether the ruler itself is straight? One "sights" along the ruler— if it deviates from the line of sight, then it is not straight: the line of sight (*the path of a ray of light*) determines whether the ruler is straight. If the path of a light ray itself is not "straight" but "curved," then space itself is curved.

Einstein concluded that gravity must be a fictitious force due simply to accelerated motion of a reference frame. This accelerated motion, he reasoned, was inertial movement through "curved" space. His general theory of relativity states that large concentrations of mass cause space near them to be curved. The motion of any object through this curved space is necessarily accelerated, and it will feel a "force" simply because of the curvature of the space.

It seems peculiar that space can be thought to be curved. It is hard to even imagine what this means. Perhaps it is best to start by considering curved two-dimensional spaces and then to return to three-dimensional (normal) space. A two-dimensional space is a surface. The surface may be flat, such as a smooth table top, or curved like the surface of a sphere. A flat, two-dimensional surface is referred to as a Euclidean surface (after the famous mathematician who developed many of the ideas of geometry). A curved surface is said to be non-Euclidean, because Euclid's theorems on geometry did not apply. On a flat surface, as Euclid noted, parallel lines never meet and the sum of the three angles in a triangle is 180 degrees. On a non-Euclidean surface, these statements are incorrect. For example, the sum of the angles in a triangle drawn on the surface of a sphere is greater than 180 degrees. Consider a triangle drawn on the Earth's surface with the base along the Equator and the two sides due-north-directed lines meeting at the North Pole. The angles between the Equator and the sides are both 90-degree angles. Thus these two angles total 180 degrees by themselves. The angle at the North Pole can be of any size from 0 to 180; thus the total sum of the angles can be from 180 to 360 degrees.

In general, a curved surface where the sum of the three angles in a triangle is greater than 180 degrees is said to have positive curvature. Curved surfaces on which the sum of the three angles in a triangle is less than 180 degrees are said to have negative curvature. (An example of a surface with negative curvature is a riding saddle.)

To get an idea as to how to find out whether our three-dimensional space is curved, let us first ask how a person confined to a two-dimensional space can tell if it is curved. We do not want to check this by climbing a tower and looking out toward the horizon or by traveling upward in a spacecraft and looking back. These methods involve moving into the third dimension to look back at the two-dimensional surface, and we cannot do the equivalent of this to look at three-dimensional space. We must confine ourselves to the two-dimensional surface and try to determine if it is flat or curved.

As already indicated, it is possible to determine if the two-dimensional surface is curved by geometry. We can, while confined to the surface, study parallel lines, triangles, and other constructions formed of (what we believe to be) straight lines. If the parallel lines eventually meet or intersect, or if triangles have angles that total other than 180 degrees, we know that we are on a curved surface. Similarly, in order to study three-dimensional space we must study straight lines and their geometry.

Actually, it is really "straight lines" that we must study. As every student in beginning geometry learns, a straight line is the shortest distance between two points. However, this is true only in a "flat" surface or three-dimensional space without curvature. On a sphere, the shortest distance between two points is a "great circle" (the line formed by intersecting the sphere with a plane passing through both points and the center of the sphere). In general, the shortest distance between two points in a given space is called a *geodesic*. It is generally considered that light will travel along the shortest path possible between two points; that is, a light ray will define the geodesic. Thus one possible way to determine if three-dimensional space is curved is to test whether light always travels in straight lines.

Perhaps the most convincing experimental evidence for the curvature of space is the deflection of light rays passing by the Sun. Einstein's prediction that mass causes space to be curved requires large mass in order to produce noticeable curvature. Thus the largest curvature of space close to the Earth would be expected near the Sun. If we accept that light rays will define the geodesics in space, then we can test Einstein's prediction if we can carefully observe light rays passing by the Sun—for example, from a distant star. The only way to do this is during a total eclipse of the Sun when the Moon passes between the Earth and the Sun and momentarily blocks out the light from the Sun.

If the light rays from different stars can be observed and compared during an eclipse, one could test whether the light rays passing near the Sun are deflected (a situation shown in Figure 6-15). If the Sun does cause space to be curved, then the

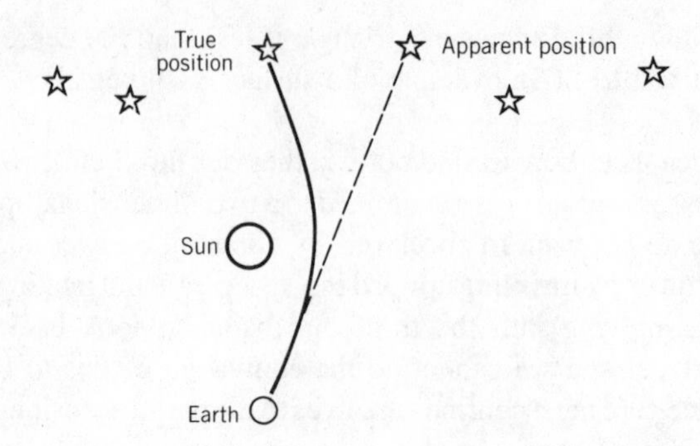

Figure 6-15. Gravitational deflection of starlight by the Sun.

starlight passing near the Sun will be deflected and the apparent position of the star, as determined by looking back from the Earth, will be shifted from its true position. The shift could be detected by carefully comparing a photograph taken of the stars observed during the eclipse with a photograph taken on another date of the same star field when visible at night (and hence when the Sun is not near the line of sight).

The prediction for the deflection of light rays near the Sun was confirmed in a famous experiment performed by a team led by the English physicist Arthur Eddington during a total eclipse of the Sun in 1919. Eddington traveled to an island off the west coast of Africa where the eclipse lasted long enough to obtain good photographs of the star field near the Sun. When he studied the photographs of the same star field taken months earlier, he discovered that the stars observed nearer to the Sun appeared to have moved as compared with stars not so near the Sun during the eclipse—and in about the amount predicted by Einstein. This experiment was quite convincing.

Because light is believed to be massless, by Newton's universal law of gravitation light should be completely unaffected by the mass of the Sun. Einstein predicted, in contrast, that mass would curve space itself and that light rays traveling along geodesics through space would be deflected by a large mass. The experimental verification of Einstein's prediction by Eddington's experiment received considerable attention in the media at that time and contributed significantly to Einstein's reputation as the new Newton of science.

Besides the deflection of starlight by the Sun, there is other experimental evidence for general relativity. In Chapter 3 (Newtonian mechanics) we discussed the fact that the mutual interaction of all the other planets with the planet Mercury causes the elliptical orbit of Mercury to rotate slowly, or precess, around the Sun. This slow precession of the orbit of Mercury has been known for some time and amounts to 574 seconds of arc per century. By about 1900, the gravitational attractions of the other planets had been calculated carefully and were known to be responsible for about 531 seconds of arc per century. The discrepancy of about 43 seconds of arc was not understood until Einstein's general theory of relativity was considered. Because the orbit of Mercury is very close to the Sun, the planet is actually traveling through a portion of space with a small but significant amount of curvature. The effects of this curved space can be shown to account for essentially all of the remaining 43 seconds of arc. This discrepancy for the precession of the orbit of Mercury is sometimes referred to as one of the "minor" problems of nineteenth-century physics. Most physicists believed the problem could be solved by some small but unimportant correction to the analysis of the motion. Instead, we see that a revolutionary new view of the nature of space is required to solve the problem.

There exist other, somewhat more complicated, experimental verifications of general relativity as well. These include the slowing of clocks in accelerated reference frames or large gravitational fields and the "red shift" of light falling in a gravitational field. It is not necessary for us to consider these other tests here, only to note that the basic principles of general relativity are now well established. At the same time, we must note that there exist several different mathematical forms of the general theory of relativity, including the form originally proposed by Einstein. The results of these theories do not differ from Einstein's version insofar as this book is concerned.

The acceptance of the need for the general theory of relativity and its prediction of curved space leads to some important specific consequences regarding the nature of our physical universe. These consequences include an overall curvature of space and the possibility of so-called black holes. If a large concentration of mass causes space near it to be curved, then there is a possibility that a sufficiently large mass may cause space to curve enough to actually fold back onto itself. This may occur either for the universe as a whole or locally near a very concentrated mass. If there exists enough mass in the universe, and if the universe is not too large, the space around the universe will eventually fold back onto itself. In such a situation, a light ray sent out in any direction would not continue outward in a straight line forever but would eventually "fall" back into the mass of the universe. The mass density of the universe would determine the actual size of the space associated with the universe. If the total mass density of the universe is not great enough, however, the curvature of space would not be sufficient to cause space eventually to fold back on itself. Such a situation would produce an infinite universe—that is, one without spatial limit.

There is an ongoing controversy whether the actual physical universe is closed (does fold back on itself) or open (never folds back on itself). Although general relativity predicts that space must be curved (because mass exists), it cannot say whether it is closed or open. The answer depends on the total mass and size of the universe, and astrophysicists find it difficult to make an accurate determination of the total mass or size of the universe.

It is known that the universe is expanding in size. This is presumably due to a great explosion, known as the big bang, which began the universe as we now know it. Astronomers can observe that light received on Earth from distant galaxies is always shifted in color toward the red part of the light spectrum, indicating large velocities directed away from the Earth.* The farthest known galaxies are believed to be about 15 billion light-years away and are receding at a tremendous rate. The question whether the universe will continue to expand forever or will eventually fall back upon itself is related to whether the universe is open or closed. If the mass density is great enough, the space of the universe is closed and the expansion of the universe eventually will stop and the universe will fall back upon itself. If the density is not great enough, the universe is open and the expansion will continue forever. There is at present insufficient evidence to draw a firm conclusion as to whether the universe is open or closed.

On a local scale, general relativity indicates that if there exists a sufficiently concentrated mass, it could curve space back upon itself in the immediate vicinity. Such a situation has become popularly known as a black hole. The mass density required is tremendous. It is greater than that in a normal star or even in a neutron star. Black holes are expected when a massive star (greater than several times the mass of the Sun) eventually burns up its nuclear fuel and collapses. When the collapsing star reaches a size of only about 18 miles, it curves space around itself enough to become a black hole. Because nothing can escape from such a folded-space object, including light rays that will follow the curved space, the object is called a black hole. Because nothing can escape from the black hole, such an object cannot be seen but has to be detected by its effects on other nearby objects. Several strong candidates for black holes are known by astronomers. They exist as partners with normal visible stars (binary systems). The normal stars can be observed to be rotating around unseen companions, and radiation (with particular characteristics) is emitted, indicating mass is being pulled from the normal stars into the black hole. Actually, the application of the ideas of quantum theory may allow a black hole to slowly emit energy in the form of electromagnetic radiation (see the article by S. Hawking listed in the references for this chapter).

General relativity indicates that the properties of space (and time) are dependent upon gravitational forces and the presence of matter. The properties of space-time are determined by light rays, which we have seen depend on electromagnetic fields. Einstein felt that these electromagnetic fields should also modify the curvature of space, and he tried to incorporate electromagnetic forces into general relativity to obtain what became known as a unified-field theory. Although Einstein and others made considerable progress in this direction, it is now clear

*This is known from the Doppler effect, which is a well-known phenomenon associated with wave phenomena, involving relative motion of source and receiver. The Doppler red shift is different from the gravitational red shift predicted by the general theory of relativity.

that a final understanding of the basic forces in nature requires the basic ideas of quantum mechanics. These ideas will be presented in the next two chapters, and we will return to the subject of unified field theories later.

Influence of Relativity Theory on Philosophy, Literature, and Art

As alluded to at the beginning of this chapter, the theory of relativity has had a significant impact on philosophy and literature and the visual arts. In some respects, it has been used to validate or support ideas that were developed anyway; in other respects it has inspired new modes of thought, although not always from a correct understanding of the theory. The idea of frame of reference in physics has been transferred to philosophy and morality as defining basic viewpoints and outlooks. Thus the physical concept of no absolute frame of reference has suggested to some individuals moral relativism, without consideration of the idea of invariant quantities, which is really an essential component of the theory of relativity.

At the same time the theory of relativity was being developed, new modes of expression were being developed in the arts and literature. Some of the Cubist painters were familiar in some fashion with the relativity theory and incorporated and interpreted its themes in their work. Poets such as William Carlos Williams and Archibald MacLeish celebrated both relativity and Einstein, and limericks were composed about the Stein family (Gertrude, Ep-, and Ein-) and relativistic time travel. Vladimir Nabokov and William Faulkner used relativistic concepts metaphorically. The role of Einstein as a Muse for the arts and literature is discussed in a recent book by Friedman and Donley, listed in the references for this chapter.

Max Planck
(American Institute of Physics Niels Bohr Library, W. F. Meggers Collection)

7

Quantum Theory and the End of Causality

You can't predict or know everything

In this chapter we discuss a physical theory that is in many ways far more revolutionary and extensive in its implications than Einstein's relativity theories. Initially, a number of shortcomings of theories discussed so far (Newtonian mechanics, Maxwellian electromagnetism, thermodynamics) will be pointed out, and then we will discuss the quantum theory. It is fair to ask why old theories should be examined if they are to be immediately discarded. Aside from the fact that one can best appreciate a new concept if one knows just how it is better than the concept it replaces, it is also true that old theories have a certain degree of validity and are often extremely useful, and many applications are based on them. In their fullest application the new theories are quite complex, and it is simply easier to use the old theories, provided their limitations are recognized.

Moreover, we have now become sufficiently sophisticated to recognize that at some future time even the latest theories or concepts may turn out to be flawed. It is therefore useful to know something about the metaphysical soul-searching scientists and philosophers have carried out over the past century or so in their efforts to achieve a better understanding of how physical knowledge is acquired and theories validated.

Scientific theories are usually expected to be logical and to make sense. Aside from the mathematical intricacies involved in a scientific theory, it is also expected to be reasonable and not violate "common sense." Of course, common sense is a subjective notion and depends very much on the range of experience of an individual or groups of individuals. Nevertheless, a new theory should not contradict major theories or ideas that are already accepted and proved, unless it can be shown that the accepted ideas contain flaws, which the new theory does not.

In evaluating a theory, one prefers to take very simple cases and verify that the predictions of the theory for them are reasonable and not contradictory. If the theory is to have any validity for a complicated case, surely it should be valid for a simple case. Similarly, the range of validity of a theory has to be tested; it must satisfy extreme cases. Indeed, it is in the tests of applicability of theories to extreme cases that their limits of validity are often determined and the need for new theories often recognized. Sometimes it is possible to introduce modifications into the existing theories and thereby handle the extreme cases.

Quantum theory, the subject of this chapter, appears to violate common sense. But quantum theory was developed to deal with very small objects (the size of atoms and molecules or even smaller) for which previous experience does not apply. Thus it is not surprising that Newtonian theory failed for such objects. Newtonian theory was developed to account for the motion of very large objects, relatively speaking—bits of dust, bullets, cannonballs, and planets, all visible to the naked eye or with ordinary microscopes—and there was no real proof that its domain of validity extended to very small objects. It is also true that Newtonian theory does not work very well at very low temperatures (approaching absolute zero), because here again the small-scale motion of matter becomes significant.

It is customary to refer to theories derived from Newtonian mechanics and from Maxwell's electromagnetism as "classical" physics, whereas Einstein's relativity theory and quantum theory are referred to as "modern" physics. (It is roughly ninety years since the inception of modern physics, but we still refer to it as modern.) As already mentioned, classical physics fails when it is used to describe phenomena taking place under extreme conditions—very high speeds or very small (atomic) dimensions, or very low or very high temperatures.

Figure 7-1 shows the domains of applicability of classical physics and the two great theories of modern physics, relativistic mechanics and quantum mechanics. Note that the horizontal scale of this figure is distorted in such a way as to give more emphasis to smaller-scale phenomena. The letters *N*, *A*, *H*, *E* on the scale denote, respectively, the size of the nucleus of an atom, an atom, human beings, and the Earth. Similarly, the vertical scale is distorted to exaggerate lower speed phenomena. The letters *So* and *c* denote the speed of sound and the speed of light in empty space. The figure should be visualized as an overlay of four regions, each smaller than the one underlying it.

Relativistic quantum theory, which encompasses the material in Chapters 6, 7, and 8, is considered the most general theory currently available. It covers almost the entire field of the figure and is valid for all dimensions and all speeds presently accessible to experiment. It is not actually known whether it is fully applicable to dimensions very much smaller than the nucleus of an atom; it is possible that it may fail for such dimensions and have to be replaced by a more

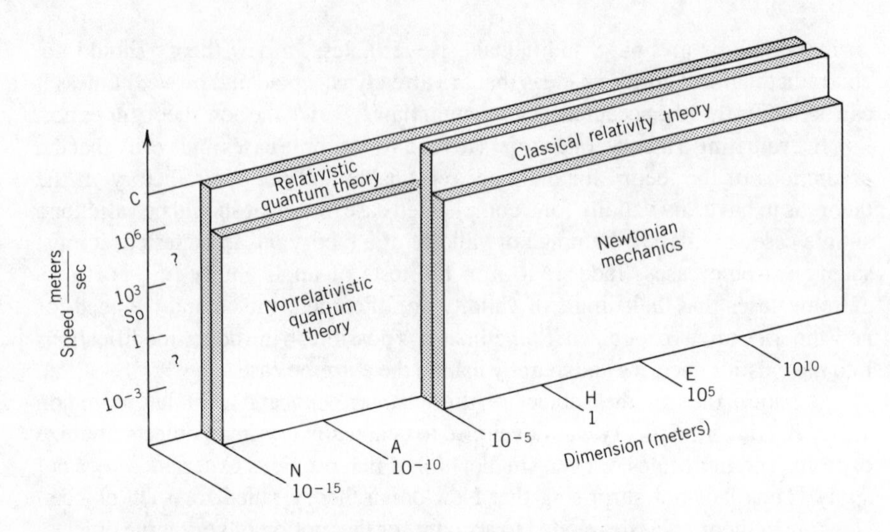

Figure 7-1. Relative domains of physical theories. *N*, typical nuclear diameter; *A*, typical atomic diameter; *H*, size of humans; *E*, Earth diameter; *So*, speed of sound; *c*, speed of light.

general theory. Relativistic quantum theory is rather difficult to use, so for many purposes it is sufficiently accurate to use nonrelativistic quantum theory, provided we are aware of its limitations. Nonrelativistic quantum theory covers a goodly portion of the region covered by relativistic quantum theory, particularly at speeds below about 10,000 or 100,000 meters/second. Classical relativity theory covers the right-hand portion of Figure 7-1—that is, objects larger than molecules at all speeds. Classical physics, designated in the figure as Newtonian mechanics, is generally more convenient and easier to use and understand than both relativity and quantum physics, and so it is used within its domain of applicability (i.e., objects that are larger than molecules and slower than a few tenths of a percent of the speed of light), even though we know it is not quite right. Table 6.2 gives some examples of how classical physics is in error; other examples are discussed in this chapter. Actually, Figure 7-1 is misleading in that the domains of applicability are not as solid and uniform as represented. There are "holes" in the overlays. Figure 7-7 represents such a "hole."

Figure 7-1 reflects the idea that all of our physical theories are but approximations of true understanding of physics. In the domains where new and old theories can both lay claim to validity, the new theory can only claim greater accuracy (the degree to which the accuracy is greater may be extremely small, even immeasurable). Thus we continue to use classical (Newtonian) physics in the design of bridges and automobiles because it is more convenient and sufficiently accurate, but for studies of the nucleus or atoms or the electronic properties of solids we must use quantum theory, because classical theory gives wrong answers in these cases. We could use quantum theory to design a bridge or an automobile, but the answers would be essentially the same as if we used classical theory.

By the end of the nineteenth century the general intellectual consensus was that basic scientific knowledge was fairly complete. Many individuals believed that the major theories of classical physics were firmly established and that possibly all that was unknown in the universe eventually could be explained on the basis of these theories. It was recognized, however, that some residual problems had to be solved.

Some of these problems arose out of the effort to apply the electromagnetic theory of light in combination with theories of matter, energy, and thermodynamics to the study of the interaction of electromagnetic radiation with matter. It was the attempt to meld these major nineteenth-century theories in a coherent manner to understand these residual problems that ultimately led to the revolutionary developments of quantum physics. It is perhaps ironic that one of the greatest achievements of nineteenth-century physics—the electromagnetic theory of light—was so flawed as to necessitate two major reformulations: Einstein's relativity theory and the quantum theory. Although Einstein also played a pivotal role in the early developments of quantum theory, no one person can be singled out as the outstanding genius who led all others in the development of the new theory. As the twentieth century developed, science became a very large-scale endeavor and scientists were more aware of each other's efforts. As a result many more individuals were in a position to make significant contributions.

We will approach quantum theory through an examination of some of these problems of nineteenth-century science and their context: (1) the blackbody radiation problem, (2) the photoelectric effect, and (3) atomic spectra and structure. These problems are listed in order of increasing importance as perceived by most physicists at the end of the nineteenth century. Indeed, the blackbody problem was considered rather minor, yet in the solution of this problem lay the seed that would lead to the full quantum theory. In fact, the three problems are intimately related to each other, and their solution became important not only for pure science but for large areas of applied science as well.

Cavity or Blackbody Radiation

It is common knowledge that metal objects change color when heated. As an iron rod is heated, for example, it eventually begins to glow dull red, then cherry red, and then bright orange or yellow. Ultimately the iron melts but, if instead of the iron a piece of tungsten wire enclosed in a vacuum or an inert atmosphere to keep it from chemically reacting with the air is used, it can be raised to quite a high temperature. The hotter it is, the more the color of the emitted light changes, so that it goes on from a bright yellow to white-hot. If the wire is enclosed in a glass bulb and the heating done electrically, we have an incandescent light bulb.

Actually, not just one color of light is emitted, but a range of colors of various intensities. The white-hot incandescent wire in the light bulb emits violet, blue, green, yellow, orange, and red light (and all colors in between) that we can see as well as "colors" of the electromagnetic spectrum that we cannot see, such as infrared and ultraviolet, as can be verified by using appropriate instruments.

Certainly, in view of the various thermodynamic concepts discussed in earlier chapters as well as the concepts of the kinetic-molecular theory of matter and the electromagnetic theory of light, it should be possible to understand the relationship between the heat put into the iron rod or tungsten wire, the temperature attained, and the range and intensity of the spectrum of electromagnetic radiation emitted. As heat energy is put into the solid, the kinetic energy of motion of the molecules and their constituent parts increases. In particular, the electrically charged parts of the atoms have greater energies of to-and-fro motion. But we know from the electromagnetic theory of light that to-and-fro (oscillatory) motion of electric charge results in the radiation of electromagnetic energy. If the oscillations are of high enough frequency, visible light will be radiated. As the temperature of the solid increases, the range of amplitudes and frequencies of the atomic or molecular oscillators increases, and the range and intensity of the emitted electromagnetic radiation increases. Thus it can be understood qualitatively how a heated body can emit light.

The next step in understanding the radiation from hot bodies consists of taking the ideas mentioned above and putting them on a more quantitative basis and making detailed experimental measurements to test the accuracy of the calculations. It was quickly realized that the amount of radiation emitted from heated bodies depends on the condition and nature of the surface of the bodies as well as on their bulk nature and temperature. Thus it is necessary to consider an "ideal" case, just as the freely falling body of Galileo and Newton is an ideal case and just as the Carnot engine represents an ideal engine.

Analysis of the situation shows that the best or ideal emitter of electromagnetic radiation at elevated temperatures is also the best absorber of radiation. A surface that can easily absorb all frequencies (colors) of light can also easily emit all frequencies. Consequently, the best emitter should have a black surface. Moreover, a white (or even more so a reflecting) surface will be both a poor absorber and a poor emitter of radiation. These facts are used by travelers in a desert: White robes reflect rather than absorb the sun's heat during the day and will retain their wearer's body heat during the night. Similarly, the use of reflecting aluminum foil on building insulation decreases heat loss during the winter and heat gain during the summer.

How does one make a truly black surface? For that matter, what is meant by a black surface? A truly black surface is one from which absolutely no incident light can escape. Any light that is used to illuminate such a surface cannot be seen. We can visualize a portion of such a surface (as shown in Figure 7-2) by considering a hollow body that has a tiny hole connecting its interior to the exterior surface. Any

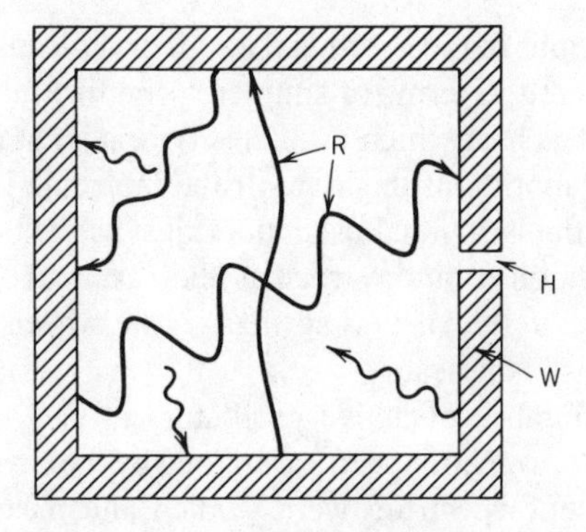

Figure 7-2. The ideal blackbody or cavity radiator. *R*, radiation of various wavelengths being exchanged by oscillators; *W*, cavity walls at high temperature; *H*, hole through which radiation is observed.

light incident on the hole will penetrate into the interior and, even though it may be reflected many times from the interior walls, it will not find its way back to the hole to escape. It is the *hole* that represents a "surface" that is the ideal absorber of radiation. On the other hand, if the interior walls of the cavity are heated, the radiation coming from the hole will correspond to the radiation emitted by the ideal emitter of radiation. Thus the ideal radiator or blackbody radiator is also called a *cavity radiator*. One could "approximate" a cavity radiator by placing a tight-fitting opaque screen in which a small hole is drilled over the opening of a fireplace. With a hot fire burning in the fireplace, the hole will be the "surface" of the blackbody.

It is possible to show that the energy emitted by an ideal cavity radiator depends only on the temperature of the radiator and is independent of any details of how the energy is generated within the cavity. Thus in the analysis of the cavity radiator, it is only necessary to assume that the atoms in the walls of the cavity are in thermal equilibrium with each other and therefore have on the average the same motional energy per atom. At any specific instant, some have more and some less energy, and the individual atoms that have more or less energy may change as time goes on, but the average energy is constant. The changes in energy of individual atoms result from interactions of the atoms with each other, either by coupling through their chemical bonds or by emitting or absorbing radiation from other atoms across the space of the cavity. In fact, the spectrum of the electromagnetic radiation emitted and absorbed—the range of frequencies and the intensity of the radiation at the various frequencies—is representative of the distribution or relative share of the total energy among the different frequencies or modes of motion. Because radiation of electromagnetic energy depends upon oscillating electric charges, it is useful to discuss the emission of electromagnetic energy in terms of atomic or molecular "oscillators" located within the individual atoms or molecules.

The possible motions of the individual atoms or molecules are quite complex, but they can be analyzed in terms of simpler springlike motions of a number of different oscillators, each of which oscillates (vibrates) at its own characteristic frequency. The total motion of the atom or the molecule is then the sum of the different motions of the individual oscillators, just as the bouncing motion of an automobile passenger on a bumpy road is the result of the up-and-down and sideways motions due to the various springs on the wheels, the bouncing of the tires, and the cushions in the seat.

Thus it is the number of active oscillators at each given frequency that determines the electromagnetic spectrum emitted by the cavity. Conversely, if the emitted electromagnetic spectrum were studied and measured and the proper mathematical analysis carried out, it is possible to deduce the manner in which the total thermal energy of the system is shared among the various oscillators at any given temperature.

This distribution of energy, or partition function as it is called, can be calculated according to the principles of thermodynamics, using the various concepts of energy and entropy discussed in the previous chapters. The results of the calculation as compared with the actual experimental measurements of the spectrum are shown in Figure 7-3, where the calculated spectrum for 7000 Kelvins is shown by a dashed line and the actual spectra for 7000, 6000, and 4000 Kelvins are shown by the solid lines. Note that in this graph, the horizontal axis is

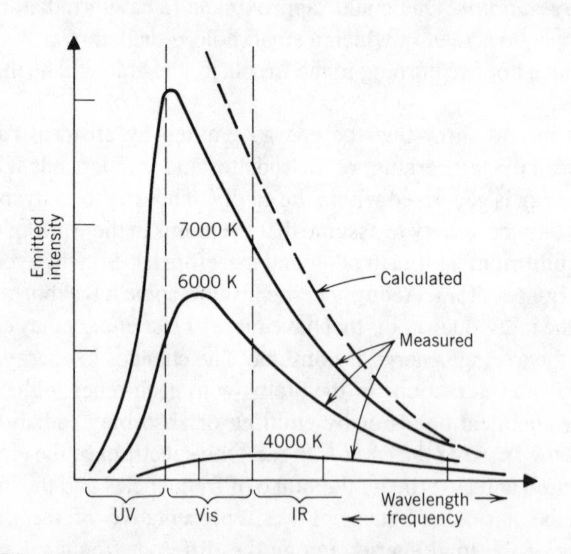

Figure 7-3. Blackbody or cavity radiation spectra. Dashed line, calculated according to classical theory for 7000 Kelvins. Solid lines, measurements at indicated temperatures. Note the spectral regions indicated: *UV*, ultraviolet; *Vis*, visible light; *IR*, infrared.

the wavelength, which is inversely proportional to frequency. Thus wavelength increases to the right and frequency increases to the left. The size of the area underneath a particular temperature graph is proportional to the total energy emitted by the cavity. As already implied, the graphs shown in Figure 7-3 depend only on temperature and are independent of specific details of atomic or molecular structure, just as the behavior of all Carnot engines is independent of their detailed design and depends only on temperature.

It is clear that the calculations agree with experiment only at long wavelengths and the disagreement becomes more pronounced as the wavelength decreases (frequency increases). In fact, at short wavelengths (high frequencies), which would correspond to the ultraviolet portion of the spectrum, the disagreement between theory and experiment is so violent as to be rather catastrophic. This disagreement became known as the *ultraviolet catastrophe*. (Of course, it was catastrophic only for the theory and those who wanted to believe the theory.) To illustrate how great the disagreement was, it became apparent that the theoretical results contradicted the principle of conservation of energy. Yet conservation of energy is one of the cornerstones of the theory of thermodynamics. So the theory even contradicted itself!

In 1899, the German theoretical physicist Max Planck published an analysis of the problem in which he modified the theory to avoid the ultraviolet catastrophe. It was this modification that started the development of the quantum theory. Planck realized that the existing theory, with its requirement that all oscillators should on the average have the same energy per oscillator, forced the ultraviolet catastrophe, because there were many more oscillators at high frequency than at low frequency. But the experimental results (Figure 7-3) clearly show that the average energy per oscillator must decrease as the frequency of the oscillator increases. This means that only a few of the available high-energy oscillators will be active. There must be some way of discriminating against the high-frequency oscillators to keep them from having the same average energy as the low-frequency oscillators. Yet the oscillators, because they reside in the atoms of the hot body, must interact and share energy with each other. So the question remains as to how the oscillators can exchange energy and still have the higher-frequency oscillators on the average with less energy per oscillator than the lower-frequency oscillators.

Planck's ingenious solution to this problem was to propose that in any situation where an oscillator gains or loses energy, its change of energy could only be in units of a certain minimum amount of energy, which he called a *quantum* of energy. Thus an oscillator could gain (or lose) one quantum of energy, or two quanta (the plural of quantum) of energy, or three quanta, or four quanta, and so on, but never a half quantum or a quarter quantum or any fraction of quantum. Moreover, each oscillator has its own characteristic size quantum, proportional to its frequency. A high-frequency oscillator would have a larger quantum than a low-frequency oscillator. He then showed that this would result in more energy being received on the average by the low-frequency oscillators than the high-frequency oscillators.

For example, if two different oscillators interact with each other and one of the oscillators happens to have a frequency that is exactly twice as high as the frequency of the other oscillator, its quantum will be twice as large as the quantum of the second oscillator. If the high-frequency oscillator "wants" to lose some energy to the low-frequency oscillator as a result of the interaction, its quantum will be equivalent to exactly two of the quanta of the low-frequency oscillator and the low-frequency oscillator will accept the energy, thereby increasing its energy by two of its own quanta. But suppose now the low-frequency oscillator "wants" to lose some energy. One of its quanta is only half the size of the required quantum for the high-frequency oscillator, and the high-frequency oscillator cannot accept the energy. The low-frequency oscillator cannot lose just one quantum to the high-frequency oscillator, so it must lose the quantum to some more compatible oscillator. Only if the low-frequency oscillator gives up two (or some multiple of two) of its quanta can the high-frequency oscillator accept the energy. As a result, the probability of the high-frequency oscillator gaining energy from a low-frequency oscillator decreases, and therefore its average energy must decrease.

We can make a fanciful analogy to this situation by imagining a community in which people interact by buying and selling various goods and services to each other. Some members of this community are willing to participate in transactions involving one or more $1 bills, others will only deal with one or more $2 bills, others will deal only with $5 bills, some only $1000 bills, and so on. If the "big spenders" want to buy a glass of milk or a pair of shoes, they must pay $1000, for example, and they get no change. On the other hand, the items they have for sale cost $1000. Unfortunately, $1000 transactions do not take place very often, and as a result the "big spenders" will not have much money left after a few transactions. The "small spenders," on the other hand, can participate in a large number of transactions and accumulate a fair amount of wealth, relatively speaking.

By incorporating this idea into the theory, Planck was able to calculate a spectral distribution function that was in exact agreement with the experimental values.

Planck was not entirely pleased with his quantum idea because it violated some of his commonsense feelings about energy. A simple example will illustrate what bothered him.

Imagine an oscillator consisting of a weight hung by a spring, as in Figure 7-4. If the weight is pulled down from its equilibrium position (say by an inch) and then released, it will oscillate up and down with a frequency that is based on the weight and on the stiffness of the spring and with an amplitude of one inch (that is, it will oscillate between two extreme positions, each of which is one inch from the equilibrium position). If it is pulled down two inches, the frequency will be the same, but the extreme positions will be two inches from the equilibrium position.

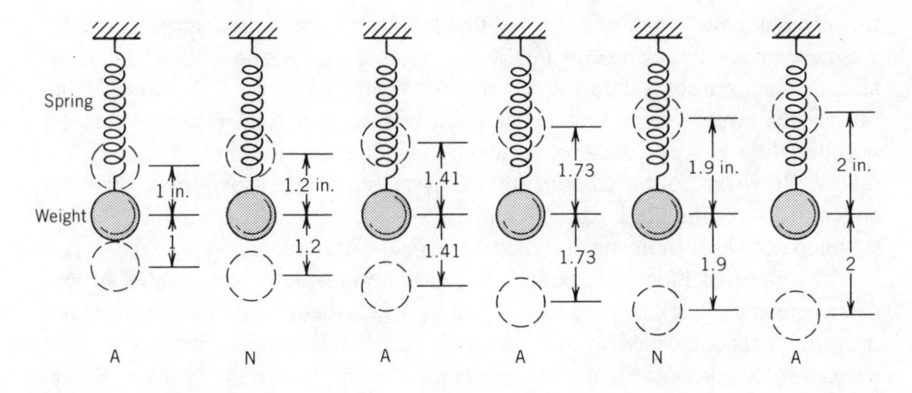

Figure 7-4. Oscillator amplitudes. *A*, allowed; *N*, not allowed.

The total energy (kinetic plus potential) associated with the oscillations in the second case will be four times as much as in the first case (the energy is proportional to the square of the amplitude of the oscillations). If the weight is pulled down 1.2 inches, then the energy should be 1.44 times as much as in the first case.

According to Planck's hypothesis, however, this is not possible, because the energy of oscillation must be exactly twice, or three times, or four times, and so on, the original amount. If the weight were pulled down a distance equal to 1.4142135... inches (the square root of 2), then the energy of oscillation will be twice the original energy and will be allowed. No amplitudes between 1 inch and 1.4142135... inches are permitted. Similarly, no amplitudes between 1.4142135... inches and 1.7320508... (square root of 3) inches are allowed, but an amplitude of exactly 1.7320508... is allowed. No amplitudes between 1.7320508... inches and 2 (square root of 4) inches are allowed, but an amplitude of 2 inches is allowed. No amplitudes between 2 inches, and so on.

Planck knew this was wrong on a large scale, and he could see no reason why it should be right on an atomic scale. In fact, he spent considerable time trying to find another way of eliminating the ultraviolet catastrophe without introducing the quantum concept, but to no avail.

Actually, it can be shown that the quantum concepts do not disagree with observed facts on the large scale. To understand this we must reexamine the relationship between the size of a quantum and the frequency of the oscillator. The size of a quantum is proportional to the frequency of the oscillator, but the proportionality constant (which is called Planck's constant) is such a tiny number (a decimal point followed by 33 zeros and then a 6 in the thirty-fourth place in the proper system of units) that the quantum of energy for the spring is a very small fraction of the total energy of the spring. As a result, the addition of one or more quanta to the large number required for a one-inch amplitude results in such a tiny

tiny change in the amplitude allowed that it is impossible to recognize that there was an even smaller change in amplitude, which would not be allowed. Thus for all "practical" purposes the allowed amplitudes vary smoothly. This is an example of the idea already expressed in the preceding section and Figure 7-1 that the quantum theory gives results for large objects indistinguishable from the classical theory. However, for the atomic oscillators, quantum theory gives quite different and correct results. It is the attempt to apply large-scale common sense to submicroscopic phenomena that causes classical theory to fail.

The value of Planck's constant, which is designated by the symbol h, was determined by Planck by a direct comparison of his theory with the experimental measurements of the cavity radiation spectrum. He received the first Nobel Prize ever awarded for work in physics because of his solution to the cavity radiation problem.

The Photoelectric Effect

The photoelectric effect was first noticed by Henrich Hertz in 1887 in the course of his experimentation to verify Maxwell's electromagnetic theory of radiation. Essentially, it is based on the fact that an electric current can be caused to traverse the empty space between two objects in a vacuum, *without a connecting wire*, when one of the objects is illuminated by light. This is shown schematically in Figure 7-5, where an electric battery is shown connected to two plates sealed in a vacuum envelope. One of the plates, called the photocathode, is connected to the negative terminal of the battery; the other plate, called the anode, is connected to the positive terminal. If, and only if, the photocathode is illuminated by light, the meter in the circuit will show that an electric current is flowing. Various experiments prove that the electric current between the two plates consists of tiny bits of matter, carrying a negative electric charge, originating in the photocathode and accelerating toward the anode. These tiny bits of matter are electrons (discovered by an English physicist, J. J. Thomson, in 1897). In the photoelectric effect they are called photoelectrons. The photoelectric effect is used in various devices such as automatic door openers, sound tracks on motion-picture films, shutter controls on cameras, burglar alarms, and numerous other devices for detecting and measuring light levels and charges in light levels.

Qualitatively, a plausible explanation of the photoelectric effect can be given in terms of the classical electromagnetic theory of light. As discussed in Chapter 6, an electromagnetic light wave consists of oscillating and propagating electric and magnetic fields. The electric field forces the electron into oscillation and gives it sufficient energy, if the amplitude of the wave is large enough, to break its chemical bond to the surface, thereby permitting the electron to be pulled out of the surface when a voltage is applied.

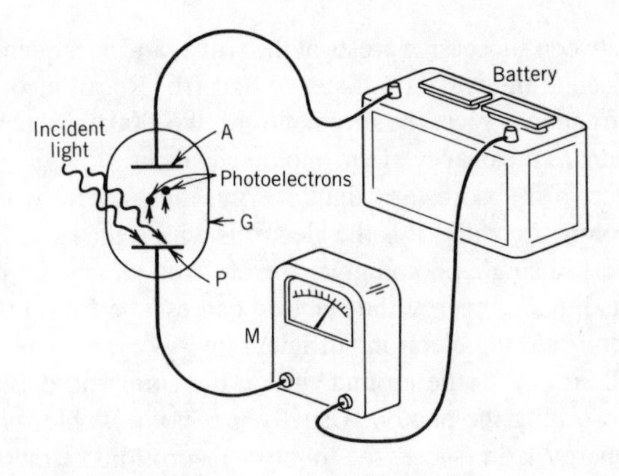

Figure 7-5. Schematic diagram of photoelectric effect. *A*, anode; *P*, photocathode; *M*, meter to indicate current; *G*, glass bulb with vacuum inside.

However, when the effect was first studied, it was not possible to explain some seemingly paradoxical results. Some colors of light, no matter how bright, did not cause the photoelectric effect when certain materials were used as photocathode, whereas other colors did cause the effect even though their intensity was much less than the first colors. For example, the most intense yellow light causes no effect from a metal such as copper, whereas ultraviolet light causes the photoelectric effect in copper no matter how weak the light. (There were also effects caused by the condition of the surface of the photocathode, and thus some experiments gave inconsistent results.)

The paradox may be visualized by using waves pounding on a beach as an analogy. As the waves come in they pick up pebbles, wood, and other debris and cast them far up on the beach. But at certain beaches, on certain days, if the distance between crests of the waves is too large, they will not move a single pebble. On the other hand, if the distance between the crests of the wave is small enough, even a tiny ripple will cast pebbles up on the beach!

Finally, Einstein, in 1905 (the year in which he published his first paper on special relativity theory as well as several other significant contributions) proposed a theory of the photoelectric effect. He called this theory heuristic because, although he could not justify it from accepted fundamental principles, it seemed to work.

Einstein proposed, borrowing and elaborating Planck's quantum hypothesis, that light energy was transported in the form of bundles or quanta of energy. At the same time, he kept the wave as the means of transport. He suggested that a quantum of light should have a special name, a *photon*. The energy in each photon of light depended upon the color of the light, in particular, on its frequency f. (Recall that according to the wave theory of light, a light wave has a wavelength λ, the distance between successive crests of the wave, and a frequency f, the number of times per second the wave oscillates to and fro. Recall also that wavelength multiplied by frequency gives the speed of light, 186,000 miles per second.) Using Planck's constant h, the energy of the photon equals hf.

The surface, if it absorbs any light energy, must accept whole photons, not pieces of a photon. Actually, it is the electrons within the surface that accept the photon energy. If a single photon gives the electron an amount of energy that is greater than its binding energy the electron can escape from the surface. If the photon does not give the electron sufficient energy, it cannot escape from the surface; it will simply "rattle around" inside the solid and dissipate the energy acquired by absorbing the photon. Usually it is not possible for the electron to store up the energy from successive photon absorptions; either it escapes or it dissipates the energy before it has a chance to absorb another photon. If the photon energy absorbed is greater than the binding energy, the excess energy will appear as kinetic energy (energy of motion) of the photoelectron. The amount of kinetic energy will be proportional to the excess photon energy.

Einstein's theory predicted that if an experiment were carried out to measure the maximum kinetic energy of the photoelectrons as they emerged from the surface, a graph of the maximum kinetic energy versus frequency of the light would be a straight line starting at some threshold frequency, which would be characteristic of the photocathode material and surface conditions. Moreover, it was predicted that the slope (or steepness) of the line in the proper units would be exactly equal to Planck's constant. Such a graph is shown in Figure 7-6. Some nine years later (in 1914) an American, Robert A. Millikan, reported the first of a series of measurements verifying Einstein's heuristic theory and showing that the pho-

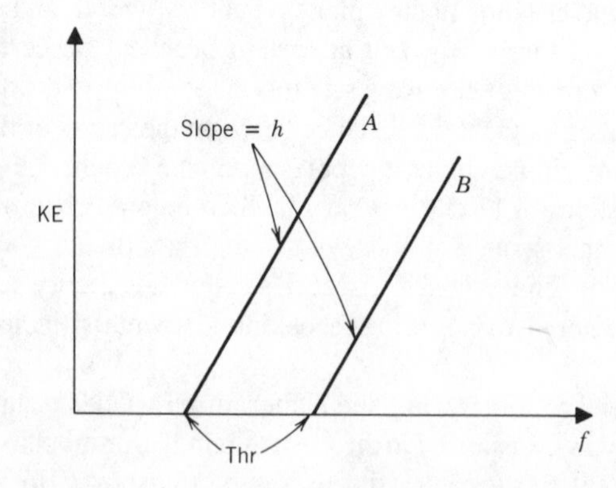

Figure 7-6. Einstein's prediction of relationship between maximum photoelectron energy and frequency. *KE*, maximum kinetic energy of photoelectrons; *f*, frequency of incident light; *A, B,* data for two different materials; *Thr*, threshold frequency.

toelectric effect could be used to measure Planck's constant h independent of the blackbody radiation problem. Einstein received a Nobel Prize for his analysis of the photoelectric effect (not for relativity theory), and Millikan also received a Nobel Prize for his experimental studies of the effect.

The photon hypothesis carried with it some implications about the nature of light that would not be expected from the electromagnetic theory. For one thing, it was necessary to consider that the photons were geometrically compact; that is, the photon travels like a bullet, not like a wave. Otherwise the energy of a single photon might be spread over a wave front several feet—even several yards—in diameter. Then when the photon was absorbed, all of its energy would have to be "slurped up" instantaneously from all parts of the wave front and concentrated at the location of the photoelectron—something not permitted by the theory of relativity because the energy would have to travel at a speed greater than the speed of light. Einstein insisted that the energy of the photon was absorbed at one particular point and therefore the photon itself had to be a very concentrated bundle of energy.

Actually, there is other evidence that the absorption of light requires the photon hypothesis. One very commonplace piece of such evidence is the granularity of underexposed photographs. If one takes a photographic negative and exposes it for the proper length of time, a positive print is obtained. If, however, the negative is exposed to only a small amount of light, the print will simply show an unrecognizable pattern of a few dots. If the exposure time in successive prints is increased by small amounts, the dots begin statistically to group themselves into the bare outlines of a recognizable image. With increasing exposure times, such a large number of dots is accumulated that the statistics become overwhelming and the final print is obtained; this is shown in Figure 7-7. Each dot represents the absorption of a photon at a specific point, confirming that the photon energy is geometrically concentrated.

A more dramatic demonstration of this "bullet" nature of light is shown by the *Compton effect*. In this effect, a beam of X-rays (known to be very short-wavelength electromagnetic radiation) is allowed to "bounce" off some electrons. It is found that the electrons recoil under bombardment by the X-rays and that the X-rays also recoil. In fact, the wavelength of the X-rays is increased (and hence the frequency is decreased) as a result of the recoil. This means that the energy of the photons decreases. It is as if any given photon had an elastic collision with an electron, just like a cue ball with a billiard ball. It is even possible to calculate the momentum of the photons (according to a formula to be discussed later) and show that in this collision, as in any elastic collision, both the total momentum and the total energy of the colliding particles (the photon and the electron) are conserved.

Still, the evidence for the wave nature of light is overwhelming: Interference effects (as demonstrated in the colors of soap bubbles), diffraction effects (apparent in the absence of sharp shadows of objects and in the behavior of diffraction gratings), polarization effects (which prove that light waves vibrate perpen-

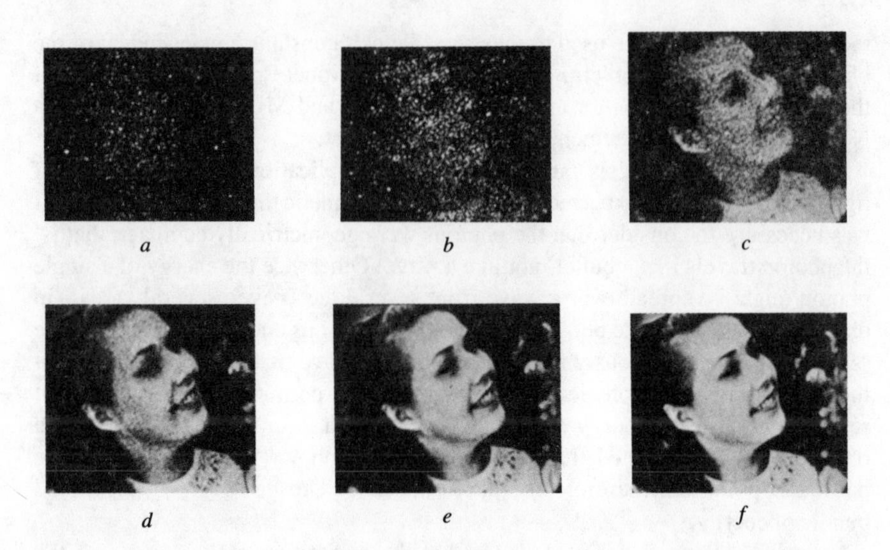

Figure 7-7. Granularity and statistics of blackening of photographic images. The number of photons involved in forming increasingly detailed reproductions of the same image. (a) 3000 photons. (b) 12,000 photons. (c) 93,000 photons. (d) 760,000 photons. (e) 3,600,000 photons. (f) 28,000,000 photons. (Courtesy of Dr. Albert Rose)

dicularly to the direction of propagation), the fact that ordinary light travels slower in glass than in vacuum, the astounding success and unifying power of Maxwell's electromagnetic theory of light were all phenomena that were well understood and precisely explained on the basis of a wave theory of light.

Thus early twentieth-century physicists were faced with a dilemma. In certain experiments, primarily those dealing with the emission and absorption of light, the photon hypothesis worked very well. In other experiments, primarily those dealing with the propagation of light (the way in which light gets from one place to another), the wave hypothesis worked very well. One wit remarked that on three alternate days of the week the experiments supported the photon theory, whereas on three other days the evidence supported the wave theory; thus it was necessary to use the seventh day of the week to pray for divine guidance.

Ultimately it was realized that light actually has a dual nature, that the properties of light can be likened to wave properties or to particle (bullet) properties depending upon the details of the particular experiment and their interpretation. The wave and particle properties are different aspects of the "true" nature of light. In fact, the wave and particle properties are intimately connected with each other. It is necessary to use the frequency f, a wave property, to calculate the energy $E\ (=hf)$ of the photon. Similarly, as will be seen later, it is necessary to use the wavelength λ to calculate the momentum, a particle property.

The Nuclear Atom and Atomic Spectra

The word *atom* literally means indivisible. In Western thought, the concepts of atomism can be traced back some twenty-five hundred years to the Greeks Leucippus and Democritus, who considered that a bit of matter could be divided into smaller bits, which could be divided into even smaller bits, which could be divided further, and so on, until ultimately there would be reached the smallest bits of matter that could not be divided any further and hence were impenetrable and indivisible. This idea was expounded further by the Roman poet Lucretius some five hundred years later in a long poetical discourse on Epicurean philosophy. In these earliest views, there were only four types of atoms, associated with the four prime substances of Aristotle. Starting with the British chemist John Dalton in 1808, it was eventually recognized that there were a number of elemental substances, each of which had its own kind of atom, perhaps distinguished by having its own peculiar shape and perhaps having attached to it a small number of "hooks and eyes" by which it could be joined to other atoms to form a molecule. A molecule is thus defined as the smallest possible bit of substance (not necessarily an elemental substance). For example, a molecule of water is composed of two hydrogen atoms and one oxygen atom. Those substances that are not elemental are now called compounds. Although compounds can be decomposed into elements, and hence molecules can be separated into constituent atoms, the elements and their atoms were initially considered as not being further decomposable or divisible, respectively.

The concept of atoms and molecules in motion led to the kinetic-molecular theory of matter and the recognition that chemistry is essentially a branch of physics. During the nineteenth century it became possible to estimate the size of atoms and molecules. Atoms are of the order of a few angstroms in diameter. (The angstrom is a very small unit of length; there are 254 million angstroms in an inch.) Many substances contain only one atom or only a few atoms per molecule; biological molecules, on the other hand, may contain several hundred or even several thousand atoms. The atoms in a molecule or a solid exert attractive forces on each other which overcome their random thermal motions and bond them together. It is possible to make a crude but useful analogy between the bonding forces and "elastic springs" connecting the various atoms with each other (the springs are used instead of the hooks and eyes mentioned earlier), but obviously the actual manner of bonding must be different.

Not too long after Dalton's work became known it was realized that matter has electrical characteristics. The final verification of the existence of the electron in 1897 confirmed that atoms themselves have an electrical structure. In fact, some 150 years earlier Boscovich, a Serbian scientist, had argued on essentially metaphysical grounds that atoms could not be the hard, impenetrable objects of the original conception but had to have some spatial structure.

Of particular significance to the realization that atoms themselves have an internal structure was the study of atomic spectra, which was developed to a high art during the nineteenth century. Under proper conditions all substances can be made to emit light of various colors. By the use of suitable devices such as prisms, this emitted light can be decomposed into its various constituent colors or spectrum, as shown in Figure 7-8.

There are various ways of causing a substance to emit its spectrum: It can be thermally excited by the application of heat, say by burning it in a flame; it can be electrically excited by passing high-voltage electricity through it when it is in the gaseous state; or it can be illuminated with other light, causing it to fluoresce. When the spectra of even the simplest substances—the elements—are studied, it is seen that they are quite complex, although in principle they can be explained in terms of the electromagnetic theory of light.

As pointed out in the discussion of cavity radiation, each atom can be considered as containing a number of electrical oscillators capable of vibrating at various characteristic frequencies. If these are set into motion by various means involving the transformation of heat or electrical energy into mechanical energy, they will radiate electromagnetic waves having a frequency corresponding to the characteristic frequency of the oscillators. The number of different colors in the spectrum of a substance, in this explanation, would correspond to the different types of oscillators present and "excited" in the atom, and the relative brightness of the colors would depend on the number of oscillators of a given type and the effectiveness with which they are excited. A study of the simplest spectrum known, that of the hydrogen atom, led to the conclusion that the structure of even the hydrogen atom must be as "complex as that of a grand piano."

At the end of the nineteenth century and the beginning of the twentieth century a number of experiments began to reveal some information about the structure of the atom. Atoms were shown to contain both positive and negative electrical charges. As already noted in connection with the photoelectric effect, the negative charges were electrons that could be "split off" from the atom. Their discoverer,

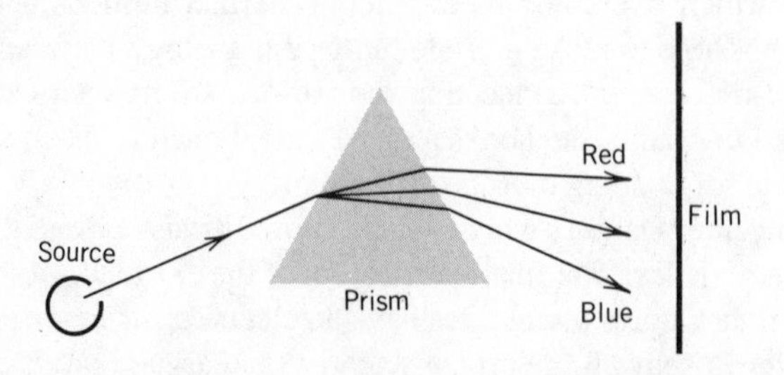

Figure 7-8. Dispersion of light into a spectrum. Light from the source is dispersed by the prism into a spectrum of colors (frequencies), which is recorded on a photographic film.

J. J. Thomson, suggested that the atom could be considered as like a blob of raisin pudding, with the electrons the raisins imbedded in the massive, positively charged "blob" of pudding. However, a New Zealander, Ernest Rutherford, working in England, then demonstrated that the atom seemed to be mostly empty space, with most of its mass concentrated at its center. Nevertheless, the chemical and kinetic-molecular properties of the atoms show that this apparently empty space is part of the size of the atom.

Rutherford then proposed the following model of the atom, which is the basis of the description given in most current popular discussions of atomic structure: Almost all of the mass of the atom is concentrated in a very tiny volume, called the nucleus, at the center of the atom. This nucleus carries with it a positive electrical charge. Outside the nucleus are a number of tiny bits of matter, the electrons, each of which carries a negative electrical charge. There are enough electrons for their total negative charge to be exactly equal to the positive charge of the nucleus. The electrons are in motion about the nucleus, traveling in elliptical orbits in the same way the planets travel in elliptical orbits about the Sun, except that the attractive force causing the elliptical orbits is electrical rather than gravitational (see Chapter 6). In other words, the atom is like a miniature solar system. The only difference between various types of atoms is in the amount of mass and charge of the nucleus and the number of orbiting electrons. In the case of the hydrogen atom, there is one electron orbiting the nucleus; for helium there are two electrons; for lithium, three electrons, and so on. Figure 7-9 shows the model for hydrogen.

There is, however, a serious shortcoming for Rutherford's model as proposed: it is unstable. An orbiting electron is necessarily an accelerating electron because it constantly changes its direction of motion. But, according to electromagnetic theory, an accelerating electron (or any accelerating electric charge) must radiate energy away, which means that the electron must eventually spiral into the nucleus.* Calculations showed that this "eventuality" would take place in

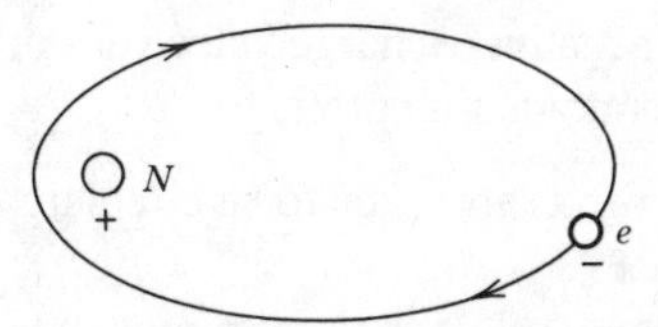

Figure 7-9. Rutherford-Bohr-Sommerfeld model of a hydrogen atom. *N*, nucleus; *e*, orbiting electron. Although the orbit is shown as highly elliptical, the nucleus is still at the center of the atom because, as shown by Sommerfeld and others, other effects cause the entire orbit to precess (rotate) about the nucleus as a center.

*In Chapter 6 we described electromagnetic waves as being caused by a vibrating or oscillating electric charge. If an orbit is viewed edge-on rather than from above, it will seem that the electron is oscillating back and forth over a distance equal to the diameter of the orbit. The essential feature of the motion for electromagnetic theory is that the velocity of the electron is changing, either because of a change in speed or a change in direction.

about 0.01 millionths of a second, accompanied by a flash of light. But this does not happen. Moreover, the calculated spectrum of the light flash, even if it were to occur, is nothing like the spectrum of light emitted by excited hydrogen atoms. Rutherford's model, as originally proposed, thus had serious flaws.

Despite the problems associated with his model, Rutherford's experimental evidence that almost all of the total mass of the atom is concentrated in the nucleus was incontrovertible. Niels Bohr, a young Danish physicist who was working with Rutherford in 1912 and who was greatly influenced by the quantum ideas of Planck and Einstein, undertook to modify Rutherford's model in the light of these ideas. Bohr published his results in 1913. He felt that the atom must be stable, and hence there must be certain electron orbits for which no radiation is emitted, despite the requirements of electromagnetic theory. Because electrons in different orbits have different total energy values, only certain particular orbits could be allowed, just as the oscillators in the blackbody radiation problem could have only certain allowed values of energy. Bohr then supposed that an atom could only change its energy by having one of its electrons change from one orbit to another orbit of different energy. Thus an atom could emit a quantity of energy—a photon for example—by having an electron "jump" from one orbit to an orbit of lower energy. Similarly, it would absorb light of a given frequency only if the photons of that frequency were of just the right energy to match the energy difference between the orbit the electron was in and some other allowed orbit, so that the electron could "jump" into the other orbit.

By a study of the experimentally measured spectrum of the hydrogen atom, and an ingenious use of the idea that when the orbital dimensions get sufficiently large, quantum ideas should give the same result as classical ideas, Bohr inferred the following postulates for the structure of an atom:

I. Most of the mass of the atom is concentrated in the positively charged nucleus. Under the influence of the attractive electrical force exerted by the nucleus, the negatively charged electrons orbit the nucleus in certain allowed stable, nonradiating orbits.

For simplicity, these orbits are taken to be circular.

II. There is a definite rule for determining which particular orbits are allowed. The angular momentum (a quantity of angular motion analogous to linear momentum for straight-line motion) of the electron, which is constant for a given orbit, must be related to Planck's constant h. Specifically, the angular momentum must be equal to a whole number n, multiplied by h and divided by 2π. (In the case of a circular orbit, the angular momentum is the mass m of the electron multiplied by its speed v and the radius r of the orbit, that is, $mvr = nh/2\pi$, where n is a whole number, 1 or 2 or 3 or 4, etc.) Thus the different allowed orbits are distinguished by the number n, which is called a *quantum number*, and $h/2\pi$ is a quantum of angular momentum. The angular momentum is said to be quantized.

III. An atom can absorb or emit radiation only in the form of photons corresponding to the energy difference between allowed orbits. The frequency f of the light absorbed or emitted must be related to the energy difference, E, of the orbits by Einstein's relation $E = hf$.

Using these postulates and the already measured values of the charge and mass of the electron and of Planck's constant h, Bohr was able to calculate the allowed energies of the hydrogen atom and thus the frequencies of the light emitted in the hydrogen spectrum.* His calculated values agreed with the measured values to within about 0.01 percent. Moreover, he predicted that there should be some particular wavelengths in the ultraviolet portion of the spectrum that had not yet been measured. These were searched for and found, just as he predicted. He also calculated from his theory the size of the hydrogen atom in its unexcited state to be about one angstrom, in agreement with experimental measurements.

Just as in the case of Newton's theory of planetary motion, further elaborations and refinements of Bohr's theory led to even better agreement between theory and experimental measurements. A German physicist, Arnold Sommerfeld, introduced elliptical orbits and corrections (required by relativity theory) to the mass of the electron because of high speed. Two Dutch physicists, Samuel A. Goudsmit and George E. Uhlenbeck, considered that the electron itself must be spinning on an axis, which should contribute magnetic effects to the calculation of the energy of the electron. Moreover, in later years this made it possible to explain the various magnetic properties of matter. All of these ideas led to the introduction of additional quantum numbers besides the quantum number n introduced by Bohr. These quantum numbers were related to the shape of the elliptical orbit, the orientation of the plane of the orbit in space, and the angular momentum of spin of the electron. Thus, in addition to only certain energy orbits being allowed, of all the possible different-shape elliptical orbits having the same energy, only certain shapes of the ellipses were allowed and only certain orientations of the plane of a

*Actually, Bohr used a much more complicated line of reasoning, but these postulates can be inferred from his work and are usually used in presenting the Bohr model. A further point worth noting is that although the energies deduced from Bohr's model are different from those assumed by Planck in his analysis of the blackbody radiation problem, Planck's analysis is still valid because the radiation emitted by the blackbody is independent of the details of the model. All that is required is that the energies of the constituent atoms or oscillators change by discrete amounts. Einstein later carried out yet a different detailed analysis of the blackbody problem in which he introduced the concept of stimulated emission of radiation. This concept is the basis on which a laser works. The results of Einstein's different analysis were the same as Planck's results.

particular shape orbit were allowed. Additionally, the electron spin was quantized to have only one allowed value of its spin angular momentum, as distinguished from its orbital angular momentum, and only two possible orientations of its axis of spin.

For all of its phenomenal successes and the seminal nature of its concepts, the Bohr theory of atomic structure had several serious shortcomings. On the one hand, Bohr's fundamental postulates were quite arbitrary and not derivable from existing theory. Yet, on the other hand, he made free use of existing classical theories. There was no justification of the assumption, for example, that the electromagnetic theory of radiation, which had been proved to work very well for orbits the size of an inch and which Bohr used in inferring his postulates, should not work for orbits the size of an angstrom. After all, radio communication, which is based on classical electromagnetic theory, is quite successful. Moreover, new quantum numbers were just introduced as needed, again without fundamental justification.

How many different kinds of quantum numbers were going to be necessary, and what rhyme or reason would be related to their appearance? In fact, the quantum number for electron spin is not even a whole number, but has the value 1/2. In addition, detailed experimental measurements of certain fine features of the spectrum led to the conclusion that, in certain circumstances, other quantum numbers might also not be whole numbers, in disagreement with the implied idea that various quantized quantities had to come in complete bundles.

Furthermore, the attempt to extend the quantitative calculations to atoms having more than one electron encountered serious snags. Calculations for the helium atom, which has two electrons, failed totally. It seemed that it would be necessary to introduce new special hypotheses and assumptions for each different kind of atom.

There were many questions to which the Bohr theory gave no answers or even inklings of answers. For example, although it could account for the specific frequencies in the spectrum of a particular atom, it said nothing about how bright or intense the light emitted at these frequencies should be. In fact, there are even a number of frequencies that should be in the spectrum according to the theory but are never observed. It turned out to be possible to deduce certain rules, called *selection rules*, to predict which frequencies are observed and which are not; but the Bohr theory gave no clue, good or bad, as to the existence of these rules. All in all, it was recognized that the Bohr theory was an incomplete or stopgap theory.

Nevertheless, the unifying power of Bohr's concepts was so great that they could be applied, if only in a qualitative or semiquantitative way, to many different areas and fields of physics and chemistry. The Bohr–Sommerfeld–Rutherford model of the atom is still discussed in all introductory high school and college courses and popular discussions dealing with atomic structure. It is the best simple illustration available, despite all its shortcomings.

Within a decade of Bohr's modification of the nuclear model of the atom, a young French physicist, Louis de Broglie, as part of his doctoral dissertation, suggested a "reason" for Bohr's postulates. Recognizing the significance of the equivalence of energy and mass proposed in Einstein's relativity theory, de Broglie suggested that because both mass and light are forms of energy, they should be describable in the same terms. Thus because Einstein had shown in his analysis of the photoelectric effect that light exhibited both wave and particle properties, so, too, matter must exhibit both wave and particle properties. Moreover, concomitant with the symmetry and intimate relationship between space and time demanded by relativity theory, there is also a symmetry and intimate relationship between energy and momentum. From relativity theory de Broglie showed that the momentum of a photon could be obtained by dividing its energy by c, the speed of light in vacuum. Thus the momentum of a photon is h multiplied by f divided by c (hf/c). However, for light waves there is a relationship between frequency f and wavelength t, namely $f = c/\mathrm{t}$. Therefore by simple algebra the momentum p of the photon must equal Planck's constant h divided by the wavelength ($p = h/\mathrm{t}$), the formula mentioned in the discussion of the photoelectric effect in the preceding section..

This relationship must hold for matter as well as for light, and in particular it must be true for the electrons in an atom. De Broglie believed that the wave nature of the electron determines how electrons get from one place to another, just as the wave nature of light determines how photons get from one place to another. Inverting the relationship between wavelength and momentum, he could use the velocity of the electron (because momentum is mass times velocity) to calculate its wavelength.

Turning his attention to the question of which orbits are allowed in an atom, de Broglie pointed out that if a whole number of wavelengths could fit exactly around the circumference of an orbit, as the electron circled around the orbit under the "guidance" of its wave nature the wave would reinforce and sustain itself in a pattern called a standing wave pattern. If, however, a whole number of wavelengths did not exactly fit the orbit, on successive circuits around the orbit the waves from the different circuits would be out of step with each other and cancel out the wave pattern entirely. Thus such an orbit could not be sustained. These ideas are illustrated in Figure 7-10.

For any circular orbit it is relatively easy to calculate the velocity of the electron in terms of the radius r of the orbit. The standing wave condition simply means that $n\mathrm{t} = 2\mathrm{t}r$, where $2\mathrm{t}r$ is the circumference of the orbit and n is a whole number. Using his relationship between wavelength and momentum, de Broglie

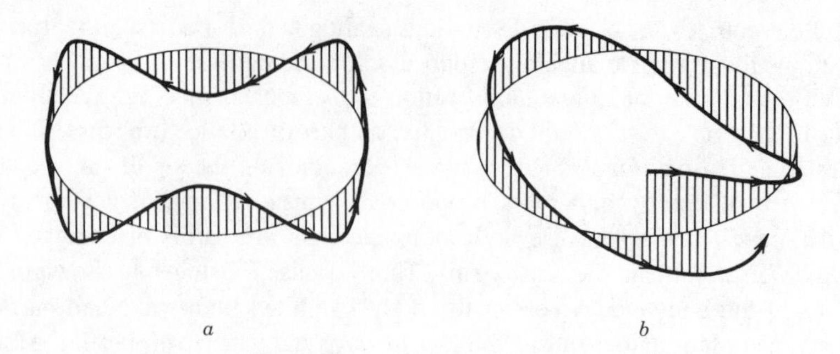

Figure 7-10. De Broglie waves for a circular orbit. (a) Reinforced waves. (b) Nonreinforcing or canceling waves.

was then able to derive by simple algebra Bohr's angular-momentum rule for the allowed orbits (Bohr's second postulate). In other words, the reason that only certain orbits are allowed in an atom is that the wave nature of the electron can establish a stable wave pattern only for certain orbits. For all other orbits the wave pattern cannot be stabilized.

Of course, the existence of a clever analogy alone does not establish a scientific principle. Fortunately, within a few years independent experimental evidence of the wave nature of the electron was found by investigators working in England, Germany, and America. (In England, the investigators included G. P. Thomson, the son of J. J. Thomson, who had discovered the electron a generation earlier.) An essential property of waves is that of diffraction, and the experimentalists showed that electrons could be diffracted in the same way X-rays can. In fact, by diffraction experiments with crystals it is possible to measure the wavelength of X-rays and electrons. Thus it became possible to measure the wavelength of a beam of electrons and show that it was exactly as would be calculated from the velocity of the beam, the mass of the electron, and de Broglie's wave hypothesis.

As has so often been demonstrated in the nineteenth and twentieth centuries, advances in one area of science often have unpredictable effects on other areas of science and human endeavor. The ability to perform diffraction experiments with electrons (and with neutrons, subatomic particles about eighteen hundred times more massive than electrons) has become a standard part of the repertoire of pure and applied science laboratories throughout the world. The electron microscope, which is found in many medical and biological research centers, is designed according to principles of wave optics and permits studies of objects that are too small to be seen with ordinary microscopes.

De Broglie's wave hypothesis, in its simple form, was of rather limited applicability to the detailed understanding of atomic structure; but it was quickly expanded and elaborated to give a complete, comprehensive, and powerful new theory of the nature of matter, capable of solving problems and answering questions with which Bohr's theory could not cope. Moreover, the ramifications and implications of this new theory have extended to almost every area of science and those areas of philosophy dealing with material knowledge. Some of these ramifications will be discussed further below.

Quantum Theory, Uncertainty, and Probability

De Broglie's significant insight was that matter, which is a form of energy just as light is a form of energy, should be describable in terms of propagating waves as well as in terms of particles moving under the influence of various forces. In the same manner as Einstein viewed light waves as determining how photons get from one place to another, matter waves should determine how particles, in particular electrons, get from one place to another. Thus instead of using Newton's laws of motion (or other principles based on Newton's laws) to calculate the motion of particles, it is necessary to use some other laws or equations to determine how waves are propagated from one place to another.

In 1926, Austrian mathematical physicist Erwin Schroedinger published a general theory for the propagation of matter waves. (This theory was an outgrowth of a seminar he had been asked to give on de Broglie's wave hypothesis.) Schroedinger's theory dealt with the propagation of waves in three dimensions, whereas de Broglie's theory was essentially a one-dimensional theory (it only considered waves traveling around the circumference of an orbit, not radially or perpendicularly to the plane of the orbit). One-dimensional waves are easily visualized: The waves set up along the strings of a violin or guitar or piano travel along the length of the string (Figure 7-11); the sound waves set up in an organ pipe or horn travel along the length of the pipe (one does not consider waves running across the pipe or horn). On the other hand, the waves set up on the surface of a drum are two-dimensional (also transverse), * as shown in Figure 7-12. Similarly, waves created by dropping pebbles in water are two-dimensional (as are the waves shown in Figure 6-5). Of course, the sound waves we hear at the concert are three-dimensional.

Matter waves are tuned by mother nature in the same way as the waves in musical instruments or auditoria are tuned by their designers. The length of the string, the diameter of the drum, the length of the organ pipe, the position of various ports on wind instruments and frets on string instruments, the dimensions

*Waves are described as transverse or longitudinal, depending upon whether the direction of the vibratory disturbance is perpendicular (transverse) to the direction of propagation of the wave or parallel (longitudinal) to the direction of propagation of the wave. Waves within the bulk of liquids or gases are longitudinal; waves in solids can be either transverse or longitudinal or a combination of both. Waves on a drumhead are transverse, as are the waves on a string. Light waves are transverse; sound waves in air are longitudinal. See Chapter 6 for a brief discussion of light waves.

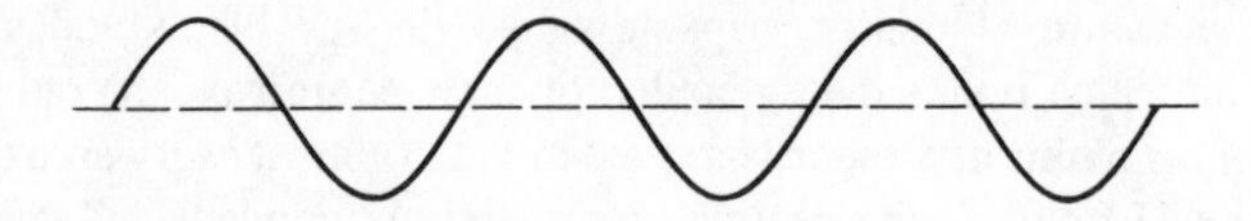

Figure 7-11. One-dimensional waves.

of the auditorium all determine certain fundamental resonant wavelengths for each system. These dimensions are all involved in *boundary conditions*, which can be specified mathematically. In addition, the propagation velocity (the speed with which the wave travels) determines the frequencies that will be resonant for the specific boundary conditions. The violinist or the piano tuner controls the speed by adjusting the tension of the string, thereby determining the resonant frequency. (The violinist also varies the boundary conditions by "fingering" the strings.)

In the case of matter waves (say for an electron) the boundary conditions are controlled by the environment in which the electron finds itself. For example, the environment for the electron inside an atom is different from the environment for an electron traveling down the length of a television picture tube. In the atom, the boundary conditions are three-dimensional, whereas in the picture tube they are one-dimensional. Moreover, the speed of the matter wave for an electron is determined by the potential energy situation in which it finds itself: Within an atom the potential energy and hence the speed of the electron is controlled by the electrical attraction of the nucleus and the electrical repulsion of all the other electrons in the atom. In a television picture tube the potential energy is determined by the electrical voltage applied to the tube. Therefore the matter waves which describe the electron will be different in the atom from what they are in the television picture tube. (This corresponds to the fact that in the Bohr model, an electron in an orbit travels a very different path than it would travel in a television picture tube—different dispositions of forces are acting on it.)

For each particular kind of wave phenomenon (vibrating strings, vibrating drumheads, sound waves, water waves, light waves, matter waves, and so on), despite all the possible variations in boundary conditions and propagation velocity, one particular governing equation permits the calculation of the wave disturbance at any point and time. This equation is called a wave equation. The wave equation has a somewhat different mathematical form for the different types of wave phenomena; that is, the mathematical form of the equation for waves on a string is different from the mathematical form of the equation for sound waves in a three-dimensional medium, which is in turn different in form from the equation for electromagnetic light waves, and so on.

A particular form of wave equation for matter waves was developed by Schroedinger; this equation is called Schroedinger's wave equation. Schroedinger's wave equation does for matter waves what Newton's laws of

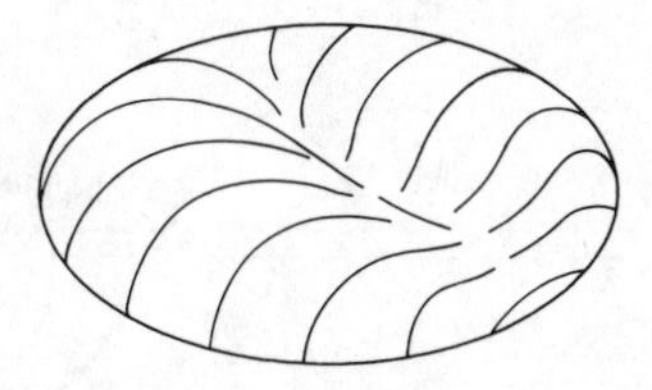

Figure 7-12. Two-dimensional waves.

motion do for particles in motion. Unfortunately, it is not possible to draw a few simple graphs to show the essential nature of Schroedinger's equation. For one thing, Schroedinger's equation and the solutions to it usually involve the use of complex numbers.* Thus the solutions that describe the matter waves are not readily shown or represented by the kinds of graphs or pictures used for other waves. The solutions, of course, are described by appropriate mathematical formulas.

The exact form of the solutions to Schroedinger's wave equation are determined, as already indicated, by various boundary conditions and the potential energy relationships for the particular situation. These solutions, then, are three-dimensional standing waves. In the case of electrons in atoms these three-dimensional standing waves define the location of the electrons within the atom, and thus they take the place of the "orbits" of the Bohr theory. In common usage they are even called orbits, even though they are conceptually quite different from the orbits of Bohr theory. Another term that is used rather than orbits is shells—we say that the electrons are found in shells around the nucleus. Even these shells are not geometrically distinct from each other; they overlap and penetrate each other somewhat. A more representative description is to say that the electrons in the atom are found in "clouds" around the nucleus.

Before discussing the results of the application of the Schroedinger theory to atomic physics, it is necessary to define some basic concepts involved in wave motion. For particle motion, the basic concepts are position, velocity, acceleration, mass, energy, and so on. For wave motion, however, some of the basic concepts are amplitude, phase, propagation velocity, and interference (see Figure 7-13). If we think of a wave as a propagating or spreading disturbance (pulsation) in a medium, or as the disturbance of an electromagnetic field, the amplitude of the wave is the maximum disturbance (at any point in space) from the equilibrium value as the wave proceeds. At that point in space, the disturbance increases from zero to the maximum value (amplitude) in one direction, decreases to zero again, reverses direction and increases to the maximum value, decreases to zero again,

*Complex numbers usually contain the square root of -1 and have some of the characteristics of two-dimensional vectors. A discussion of complex numbers and their significance would be too large a digression from the principal goals of this introduction to quantum physics. Suffice it to say that they are extremely useful for proper mathematical description of many phenomena.

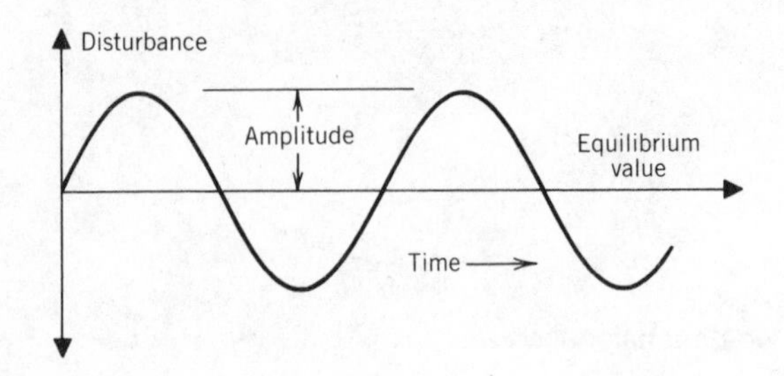

Figure 7-13. Characteristics of waves.

reverses back to the original direction, increases to the maximum value, and so on; and, if it is a sustained steady wave, the cycle of disturbance at a point waxes and wanes repeatedly. At any instant, the disturbance may be in a maximum phase or a minimum phase or some phase in between.

The *phase* of the wave refers to what part of the disturbance is occurring at the particular point in space at a particular instant of time. At other points in space the disturbance will also go through a cycle, but not necessarily at the same times as at the first point. Actually, all the various points in space affected by the wave act like the simple harmonic oscillators discussed above; but at any one instant of time, at different successive points along the wave the phases of the oscillators vary from zero disturbance to maximum in one direction, back to zero, and then to maximum in the opposite direction, back to zero again, and so on. As time goes on, the phases change in a progressive manner in the direction of propagation and the speed with which the changes progress is called the *propagation velocity*.

Sometimes more than one wave of disturbance is present in a medium at the same time, with a resulting total disturbance that can be quite large, depending on circumstances. If the circumstances are correct, the disturbances caused by the individual waves alone are added to each other when they are all present. If these disturbances have the same wavelength and frequency, then an interesting phenomenon called interference takes place. The resulting disturbance will also be a wave of the same wavelength and frequency, whose amplitude and phase will depend upon the amplitudes and relative phases of the individual waves (Figure 7-14). If all the individual waves have the same phase at a particular point at a given instant of time, then the resulting wave will have an amplitude equal to the sum of the amplitudes of the individual waves, and hence will be quite large. This is called *constructive interference*.

On the other hand, it is also possible for the individual waves to have such phase relations that at any given point, when one wave disturbance is at its maximum amptitude in a given direction, another wave may be at its maximum amplitude in the opposite direction, with the result that the two waves cancel each other completely and there is no net disturbance at all, just as if there were no wave

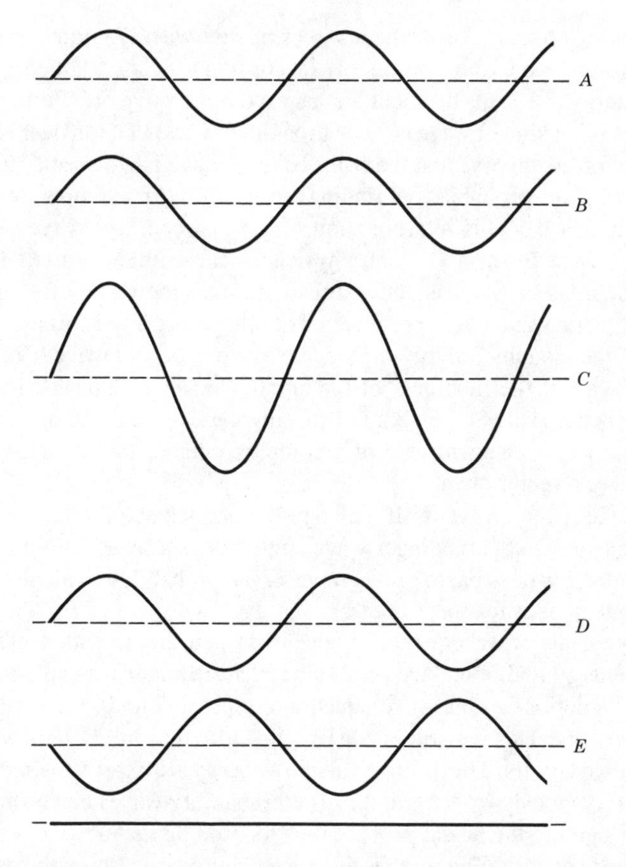

Figure 7-14. Interference of waves. Wave A plus wave B interfere constructively to give wave C. Wave D plus wave E interfere destructively to give no wave at all.

present whatsoever. This is called *destructive interference*. These are the two extreme cases; there will be various degrees of constructive and destructive interference between the two extremes. Figures 7-13 and 7-14 show these quantities and concepts for waves involving real numbers; the ideas are similar for the complex numbers involved in matter waves, but not as easy to depict. Interference effects are used to explain the appearance of colors in soap bubbles and oil films on water on one hand and the operation of the Michelson interferometer on the other hand.

The energy transmitted by a wave disturbance is related to the amplitude of the wave. The rate of transport of energy by the wave is not directly propotional to the amplitude of the wave, however, but rather to the square of the amplitude. This is true for light waves and sound waves.

Returning now to Schroedinger waves (matter waves) a number of questions arise: Does the wave transport energy (according to relativity theory matter is a form of energy)? What is the medium that carries the wave or, alternatively, what is waving? What kind of wave is the Schroedinger wave? The answer is (and this is a postulate of the theory) that the Schroedinger wave is a probability wave! The wave "carries" the probability of finding the electron (for example) at the particular point in space. (This interpretation of the Schroedinger wave was first put forward by Max Born, a German physicist. Schroedinger himself felt that the electron in an atom was somehow spread out over the wave.) By analogy with electromagnetic waves, it is postulated that this probability is proportional to the "square of the modulus" of the amplitude of the probability wave. (It is necessary to say "square of the modulus" of the amplitude rather than just square of the amplitude because the Schroedinger wave involves complex numbers; the modulus is a particular characteristic of a complex number that is analogous to the magnitude of a vector quantity.)

Even though the wave itself may have a complex amplitude, the probability calculated from the Schroedinger wave amplitude is a "real" (ordinary) number, and therefore it is possible to draw graphs of the probabilities calculated from the Schroedinger waves for specific cases.

Applying his wave equation to the hydrogen atom, with the appropriate potential energy and boundary conditions, Schroedinger was able to show that acceptable solutions of his wave equation are possible only for certain discrete values of energy. This is similar to de Broglie's idea that only certain waves would fit into allowed orbits. The discrete values of energy are exactly the same values as calculated previously by Niels Bohr. The probability waves can be calculated in detail, thus giving the actual probabilities for finding the electron when it has a given value of energy. The result is that for a given energy, the associated electron orbit is not a sharply defined circle or ellipse but rather a "cloud" of "probability density" extending over a fair-size region of space (Figure 7-15). Where the cloud is most dense, the electron is most likely to be found, and where it is least dense, the electron is least likely to be found. As it turns out, the cloud is always fairly dense at a distance from the nucleus equal to the radius calculated from the Bohr theory. This should be expected because the Bohr theory worked fairly well for the hydrogen atom.

What is quite different from the Bohr theory is the extent of the cloud. For example, there is a small probability that the electron could be found right at the nucleus. Even more surprising, there is a very small probability that the electron could be found a mile away from the nucleus. It is a very small probability—much less than one chance in a billion years—but it is, in principle, mathematically not zero, even though for all practical purposes it is essentially zero. Moreover, as shown in Figure 7-15, the clouds associated with the various possible energy levels often have more than one region where the probability is greater than in the immediately adjacent regions.

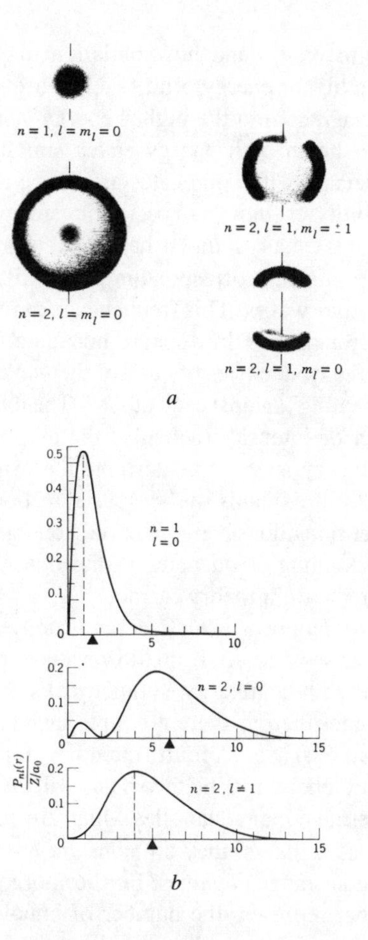

Figure 7-15. Probability density for a hydrogen atom. (a) Three-dimensional. Each figure represents a different "orbit," having different values of the various quantum numbers. (b) Graph of total probability density at a given radius for the same "orbits" as above. The dotted lines show the radii of the corresponding Bohr orbits. (Reproduced by permission from Robert Eisberg and Robert Resnick, *Quantum Physics*. New York: John Wiley & Sons, 1974)

The Schroedinger picture also makes it possible to solve the problem of the stability of the orbits, which Bohr had to handle by the first of his special postulates. Using the probability waves, it is possible to calculate the probability of finding an oscillating or accelerating electric charge in the atom when an electron is at a given energy level. The result of the calculation is that there is zero net oscillating electric charge, and therefore zero radiation of electromagnetic energy. Thus the orbit is energetically stable.

It is also possible to understand how radiation may be emitted when an electron changes from a higher energy orbit to a lower energy orbit. During the time the electron is changing from the higher energy state to the lower energy state, it can be said to be in both energy states simultaneously; that is, the probability waves associated with both states are active and interfere with each other. When the probability of finding an oscillating electric charge is calculated under these conditions, it is usually found to have a value greater than zero, but the value oscillates with a frequency corresponding to the difference of the frequencies of the two simultaneous waves. This frequency is exactly the same frequency as would be calculated from Bohr's third special postulate. The calculation simply expresses the idea that the frequencies of each of the waves corresponding to the two energy states are "beating" against each other.* The mathematics also permits calculating the strength or intensity (actually, the amplitude) of the electrical oscillations. Thus the theory is able to determine the brightness of the spectral lines emitted. Moreover, it explains the various selection rules that have been found to govern whether transitions between certain energy levels even occur and contribute their corresponding frequencies to the observed spectrum. This is simply beyond anything the Bohr theory can do.

The various quantum numbers, which were introduced on a rather arbitrary basis in the old theory, arise naturally from the various requirements imposed on the boundary conditions and the inherent geometrical symmetry of the problem. In the simple Schroedinger theory, there are three quantum numbers associated with the three dimensions of space. (A fourth quantum number, usually associated with electron spin, is not inherent in the theory, as will be discussed below.) The quantum numbers represent certain quantities which are conserved in a particular problem and are quantized; that is, they are constant but can have only discrete values and not a continuous range of values. Furthermore, the quantum numbers, although integers, do not represent the number of bundles or units of the conserved quantity but are used to calculate the discrete values of the conserved quantity. In the case of the simple hydrogen atom, these quantities are the total energy of the atom, the magnitude of the angular momentum of the electron, and one component of the vector which represents the angular momentum.** It is

*The phenomenon of "beats" between waves of slightly different frequencies is most easily demonstrated by simultaneously sounding two tuning forks of very slightly different frequencies. The resulting sound will oscillate in loudness with a beat frequency equal to the difference between the frequencies of the two tuning forks. A similar effect is noted in poorly tuned radios that happen to be receiving two stations of slightly different broadcast frequencies simultaneously. A very loud annoying whistle is heard, and the frequency of the whistle is the difference between the frequencies of the two stations. The same principle is used in tuners for most radios and television sets, where it is called heterodyning.

**Angular momentum, like ordinary linear momentum, is a vector quantity. As may be inferred from Fig. 3-7 and the discussion of vectors in Chapter 3, any three-dimensional vector can be considered to be the sum of three mutually perpendicular vectors, which are called components of the original vector.

even possible to show that for the waves with high quantum numbers, the regions of high probability begin to look increasingly like orbits in the classical sense. This illustrates the idea that quantum physics gives essentially the same results as classical physics in the domain where classical physics has been proven valid, as discussed at the beginning of this chapter.

Although they are not inherent in the Schroedinger wave mechanical theory as originally formulated, it is also possible to graft onto the theory corrections for magnetic effects on the energy due to the spinning electron and thus make the refined calculations that were not possible for the Bohr theory and its elaborations by Sommerfeld and others. Moreover, mathematical techniques have been developed to apply the Schroedinger theory to all atoms and to achieve excellent quantitative agreement with experiment. This was an area where calculations based on the Bohr model had failed miserably. (Recently it has been shown how to make the Bohr theory apply to some of these cases; however, this was done only after the Schroedinger theory showed what the correct answer should be.)

The Schroedinger theory, although powerful and useful, does have certain limitations. Unlike the Bohr theory, however, it is possible to work around these limitations. The chief limitation, from a fundamental and conceptual viewpoint, is that Schroedinger's wave equation does not satisfy the requirement, imposed by the theory of relativity, that the form of the equation be the same for all observers. Attempts to modify the Schroedinger theory to meet this requirement have met with limited success. In fact, it is for this reason that the Schroedinger theory does not itself satisfactorily treat the problem of the spinning electron and the various magnetic effects associated with the spinning electron. There is no basis in the Schroedinger theory for determining the electron spin and its associated quantum number.

Within a few years of Schroedinger's work, P. A. M. Dirac, an English mathematical physicist, published a new theory, called a relativistic quantum theory, that incorporated the demands of relativity theory from the very beginning. In this theory no waves are involved; indeed, there are no quantities involved that would permit drawing simple pictures or models of an atom. In the theory, the idea of electron spin (the magnetic effects that are pictorially ascribed to electron spin) turn out to be consequences of the requirements of relativity, just as magnetic effects in general can be attributed to relative motion of electric charges. Dirac's theory yields all the quantum numbers of Schroedinger theory and additionally the half-integer quantum number associated with electron spin. It becomes quite clear why there are four quantum numbers associated with an electron: boundary conditions and mathematical symmetries must be associated with the four dimensions of space-time. The Dirac theory also predicted that in addition to the existence of the negatively charged electron, there should also exist a positively charged particle having the same mass as the electron. This particle, now called a positron, was discovered some four years after Dirac predicted its existence.

The Dirac theory is not convenient to apply, and for most purposes the Shroedinger theory can be readily modified to allow for the effects of electron spin. Thus the Schroedinger theory is very useful for solving problems throughout the domain of atomic, molecular, and solid state physics.

One significant aspect of a probability wave is that we cannot talk about finding an electron at a particular point in space because we cannot localize a wave at one particular point in space. A wave is necessarily spread out, and thus the probability of finding an electron at one point is also spread out. We say that the electron is "smeared out." This idea has some interesting implications. The introduction of the idea of probability waves leads to a number of interesting implications. The introduction of probability into the discussion of a physical situation means that there is some uncertainty. Even if something is very highly probable, it is by definition necessarily slightly uncertain. Often when something is said to be probable or uncertain, the statement is made only because there is not enough time or a good enough measuring instrument available to make a determination with certainty. In quantum physics, however, it is believed that the uncertainty is inherent in the nature of things and cannot be removed regardless of how well the measurements may be made. Simply stated, it means that if predictions are made based on past events and a complete, well-understood theory, the results may not be certain. Thus the direct and rigid connection between cause and effect is destroyed, because it is not certain that the cause will lead to the exact effect.

To see how this occurs, it is necessary to discuss what is involved in making or specifying a precision measurement. Suppose that for some reason it were necessary to measure the diameter of a round object such as a basketball. Suppose further that it is assumed the surface of the ball is perfectly smooth and perfectly round so that there need be no concern about superficial irregularities. The ball might be set up against a suitable measuring stick, and the person making the measurement would try to look perpendicular to the measuring stick to see exactly where the projection of the extreme edges of the ball would fall on the scale. If this person has a fairly good eye for such things, the measurement might come out to be 12½ inches. The question might then be asked: Is it exactly 12.5 inches, or is it possibly 12.499 inches or maybe 12.501 inches?

If the question is sufficiently important, the person making the measurement might get a large machinist's micrometer caliper, adjust it with the right "touch" so that the jaws of the caliper just make contact with the ball (but do not squeeze it), and determine that the diameter of the ball is 12.500 inches. Again the question might arise: Is it possibly 12.4999 or 12.5001? The person making the measurements might then determine that the precision of the micrometer caliper is at best .0001 inch, and thus the last question is not answerable with the instrument at hand. The error of measurement is possibly as much as .0001 inch. It might be only

.00001 inch, but this cannot be proved with either instrument or sense of touch. In adjusting the instrument to fit snugly against the ball, the measurer might inadvertently squeeze it out of shape by .0001 inch, particularly if it is a soft ball.

Two conclusions may be drawn from the foregoing discussion. One is that, at least in the physical world, it is not possible to know something such as the diameter of a ball unless it can be measured. In other words, knowledge must be based on experimental observation or measurement. This statement is at the core of the philosophical school called logical positivism. The other conclusion is that the attempt to measure an object or a situation may disturb or distort the thing that is to be measured. (In the case of the soft baseball, the distortion was only about .0001 inch, but this set a limit to the precision to which the diameter of the ball could be measured and hence a limit to the precision to which the diameter could even be known.)

An attempt to measure something that distorts what is being measured is not uncommon. For example, it is often observed in studies of the effectiveness of various medications for combating illness that the test patients who take these medications often report they feel better whether or not the medicine is really effective. The fact that they are being observed changes their medical response. It become necessary to give some of the patients a placebo (a substance that has no effect at all) without advising them whether they are receiving the placebo or the true medication. The idea is to observe patients without their knowledge. If the observer is clever enough, a way can be devised so that the patient's subjective reactions will not distort or interfere with the observations.

In considering observations on matter, then, the question arises as to whether it is always possible to find a way to make a measurement to any precision desired without disturbing the experimental situation. It may be very difficult to do and may involve more effort than the knowledge is worth, but it is a matter of principle that needs to be understood. Suppose, for example, that it was desired to carry out an experiment to observe an electron and to study the probability waves described in the Schroedinger theory. This would be a very significant experiment. Schroedinger felt that the solutions to his equation did not merely give a probability of finding the electron at some point in space but actually meant that the electron was itself smeared out throughout the region of the electron cloud. Others felt that it should be possible to see the electron, albeit with only a probability, at specific points in space. Maybe it even would be possible to see the electron actually spinning on an axis.

How does one see an object as small as an electron or even an atom? One, in effect, looks at it with light. But a wavelength of the light with which we actually see is about five thousand times larger than the diameter of an atom. In the same sense that an ocean wave sweeping over a small pebble is not disturbed by the presence of the pebble, a light wave sweeping over an atom would not be noticeably affected by the presence of the atom, and we cannot see any effect of the presence of the atom on the light wave. The wavelength is too long (see Figure 6-5). One should use "light" with a wavelength which is as small as the atom or electron; in fact, the wavelength should be much smaller. Very short-wavelength electromagnetic radiation—X-rays and even gamma rays—should be used to "see" the electron.

But now the dual nature of electromagnetic radiation comes into play. The X-ray or gamma-ray photons that are guided by their waves have a considerable amount of energy and momentum, according to the ideas of Einstein and de Broglie. In fact, their energy and momentum are perhaps a thousand times greater than that of the electron being observed. As soon as one of the photons interacts with an electron, the collision will be so violent as to knock the electron completely out of its orbit and thus the attempt to measure the position of the electron with great precision totally disturbs the situation. It becomes necessary to make the measurement with longer-wavelength photons to avoid disturbing the electron unduly, and therefore the precision of the measure is not as good as desired.

According to the quantum mechanical theory of measurement, the act of observation of the electron will force it into a particular quantum state it would not otherwise have entered. Before the measurement it was not known in which orbit the electron could be found; the measurement actually "put" it into some orbit.

These ideas were first expressed by Werner Heisenberg, a German physicist, in the form of a principle called the *Heisenberg uncertainty principle* or the *Heinsenberg principle of indeterminacy*. In words, the principle states that *it is impossible to measure, predict, or know both the position and momentum simultaneously of a particle, with unlimited precision in both quantities*. In fact, the error (or uncertainty) in the position multiplied by the error in the momentum must always be greater than Planck's constant h divided by 2π. For example, if the error in one of the quantities is numerically smaller than about 0.01 h, then the error in the other quantity must be greater than $100/2\pi$, or about 16. Note that Heisenberg's principle applies to quantities in specific related pairs (the position in a given direction and the momentum in that direction). It also applies to the energy of the particle and the time at which it has that energy. It does not apply to a simultaneous measurement of the momentum and energy of the particle. Generally speaking, in most measurement situations, the uncertainty required by Heisenberg's principle is quite small and beyond the reach of the available measuring instruments; however, there are certain situations in which the principle plays a useful role. The attempt to measure the position of an electron in an atomic orbit, discussed above, is one such situation.

The principle has had a profound effect on discussions of metaphysics and fundamental concepts of knowledge. It destroys the idea that the universe is completely determined by its past history. This idea was strongly suggested by Laplace almost two hundred years ago as a result of the success of Newtonian mechanics. Laplace stated that if the positions and velocities of all the bits of

Werner Heisenberg and Niels Bohr
(Photo by Paul Ehrenfest, Jr., American Institute of Physics Niels Bohr Library)

matter in the universe were known at one time, and if all the various force laws were known, the positions and velocities of all these bits of matter could be calculated and predicted for any future time. All future effects would be the result of earlier causes. Even if the task of measuring all these positions and velocities were humanly impossible, and even if the discovery of all the appropriate laws were impossible, nevertheless the positions and velocities did exist at a previous time and the laws do exist; therefore the future is predetermined.

But Heisenberg's uncertainty principle says this is not so. It is, in principle, impossible to make the measurements with sufficient precision or even to calculate from them the future positions and velocities because we cannot know the future positions and velocities. (According to the positivist mode of thought, if we cannot measure, we cannot know or predict, nor can nature know or predict.) There are limitations on causality, but this does not mean that the future is completely unknown. We can calculate probabilities that things will occur in the future, and statistically these calculations will be borne out. It is only for the individual electron that we cannot predict.

Not all physicists accepted the uncertainty principle and its consequences. Einstein, for one, did not like it, and he had a long series of arguments with Bohr and others about its validity. He made numerous attempts to refute the principle or to find examples where it would lead to obvious error or paradox. Ultimately Einstein conceded that predictions based on it were valid but insisted that there must be some more satisfying principle that would account for the results of quantum theory but would still preserve causality completely. His primary argument was a philosophical intuition that "God does not play dice with the universe." Nevertheless the prevailing mode of thought among most physicists today is that the uncertainty principle is valid and useful and there are indeed limitations on causality.

The Use of Models in Describing Nature

Almost all of the preceding discussion of the nature of matter and energy has been based on the use of models. The models supposedly represent the "true reality" in the sense of Plato's Allegory of the Cave. This has been characteristic of the historical development of physics. The caloric model of heat described it as a colorless and weightless fluid. Bohr's model described the atom as a miniature solar system. In discussing the nature of an electron, it was thought for a while to be like a very tiny spherical ball that could spin on its axis; later it was considered to be like a wave. Light is considered as exhibiting a dual nature.

It may well be asked why these models should be made or even why old or flawed models should be discussed when it is known that they are not correct. Perhaps we should not try to find the "true reality"—whatever we observe or whatever exists is real, and perhaps there is nothing else underlying existence. There is, however, a very practical reason for making models: They are convenient and make it possible to sum up in a relatively few coherent words an intricate collection of physical phenomena. They make it possible to assimilate and integrate into the human mind new facts and knowledge, and relate them to previous knowledge. When an atom is said to be like a miniature solar system, a picture immediately springs to mind which shows that an atom is mostly empty space, that it should be possible to remove one or more electrons, and so on. Thus the model serves not only as an analogy but also makes it possible to synthesize (in our minds) some other possible properties of atoms that we might otherwise not have guessed.

But is the model real? After all, if a hobbyist builds a model airplane to be set on a shelf in a room, it is definitely not the real thing. It is only a representation of the real thing, and we must keep this in mind in our examination of the model. For example, we must not think that the engines are made of plastic; or if it is a flying model, we must think that the engines used in the model are like the real engines or even provide the model with the same flight characteristics as the real object. Many pitfalls have led physicists astray (to mix metaphors) when models are used. These have led to a number of paradoxes and attempts to construct rival models for the same set of phenomena.

Although recognizing the claimed utility of models for practical purposes, many physicists have argued that the sooner we can abandon the use of pictorial models the sooner we can obtain a profound understanding of physical phenomena. After all, a model is not useful if it cannot be treated mathematically to see if it will lead to a detailed quantitative account of the experimental data. The model must be expressible in terms of equations. That being the case, why not abandon the model entirely and just write down the basic mathematical assumptions that lead to the equations? All that is necessary is that the assumptions not be inconsistent with each other and that there not be too many of them. Then we need not be concerned about whether these assumptions make sense in terms of some simple-minded model.

Thus the model now becomes a set of equations, and there should be no question as to whether the model is real or not. An atom is not a set of equations.* In this vein, various economists construct "models" of the world economic system and make predictions—usually dire—about what will happen if certain trends continue.

At about the same time Schroedinger was developing his wave mechanical model, Heisenberg and others were developing strictly mathematical models of atomic phenomena. Heisenberg felt that the Rutherford–Bohr–Sommerfeld model was too full of inconsistencies and that a model of the atom should make direct use of the experimental observations, taken primarily (but not entirely) from studies of the spectra emitted by the atoms under various excitation conditions. He made use of the properties of mathematical quantities called matrices and his theory was therefore called matrix mechanics. He used different assumptions than Schroedinger—he did not like Schroedinger's theory. Schroedinger, on the other hand, found Heisenberg's assumptions "repellant." Interestingly enough, very quickly Schroedinger (and others as well) was able to show that his theory and Heisenberg's theory were mathematically equivalent! Schroedinger started out with the idea that physical phenomena were by nature continuous and that under certain circumstances (with proper boundary conditions) quantization and quanta appeared. Heisenberg, however, started out with the assumption of quantization of phenomena at the fundamental level. He also incorporated into his assumptions a formulation of his uncertainty principle. Quite often in detailed calculations both Schroedinger's and Heisenberg's approaches are used together, because Schroedinger's theory can be used to calculate certain quantities needed in Heisenberg's theory.

An even more abstract formulation was used by Dirac when he developed his relativistic quantum mechanics. This formulation is consistent with the use of both wave mechanics and matrix mechanics.

---

*Interestingly enough, a century ago there was a significant school of thought among physicists which argued that atoms were not real, they were only mental constructs—they could not be seen with any instruments, and their existence was only inferred—and thus one should not talk about the kinetic-molecular model of matter.

At the beginning of this chapter we noted that it is often conceded that there might be yet a further development of an overarching theory which would be more general than relativistic quantum mechanics. It is sometimes speculated, or even hoped, that this new theory will contain within it new underlying concepts or new variables of which we are currently not aware. These are sometimes called hidden variables. Those who are uncomfortable with the uncertainty principle, even though they are currently compelled to concede its apparent validity, believe that these hidden variables will once again restore complete causality to physics. There are indeed new theoretical developments that seem to hold the possibility of uniting not only relativity theory and quantum theory but also gravitational theory and the theory of various forces active within the nucleus of the atom as well. Some of these are discussed in the next chapter although the desired hidden variables are not included in these developments.

The Impact of Quantum Theory on Philosophy and Literature

Quantum theory apparently sets limits on the rigorous causality associated with classical Newtonian physics. The Heisenberg principle is viewed by some writers as making possible, on the most fundamental level of the material universe, the concept of free will, because free-will decisions are necessarily unpredictable. It is not that an individual electron will "choose" to behave in an unpredictable manner but rather that nonphysical influences (for example, the human will or divine intervention) may affect the behavior of the material universe. In this respect free-will decisions are indistinguishable from random decisions. However, it must be emphasized that while the specific behavior of individual atoms and molecules may not be predictable, the overall statistics resulting from the astronomical number of atoms and molecules that make up a simple biological cell is so overwhelming that the average behavior of the very large number of atoms in a biological cell is highly predictable; and significant deviations will occur only very rarely. Thus it is not at all clear that we should believe that a universe governed according to modern physics allows any more freedom of choice than one governed according to classical physics.

Some writers have interpreted the uncertainty principle as demonstrating that there are limitations to what can be known in a material sense, saying that we can only claim knowledge about things that can be measured. Things which in principle cannot be measured are therefore unknowable and therefore nonexistent in a material sense.

As can be readily imagined, many thinkers take umbrage at such assertions as lack of causality or limitations on knowledge. Thus there has been a continuing lively debate, almost from the very inception of quantum theory, among physicists, philosophers, and philosophers of science on these matters, which are at the very root of basic concepts about knowledge. Philosophers tend to regard physicists as rather naive about such matters, and physicists tend to regard philosophers as somewhat out of touch with physical reality. However, as already mentioned, there is some disagreement among physicists themselves as to how quantum theory should be interpreted. A detailed annotated bibliography and guide to some of the numerous articles and books on the "proper" interpretation of quantum theory and its metaphysical-philosophical foundations and implications is given in the article by DeWitt and Graham listed in the references for this chapter.

The prevailing view among physicists as to the significance and interpretation of quantum theory was developed by a group of physicists called the Copenhagen School, among the most famous of whose members were Neils Bohr and Werner Heisenberg. An example of their mode of thought is the *complementarity principle*, which was introduced by Bohr in 1928 in order to present the Heisenberg uncertainty principle in more general terms. The complementarity principle states that on the scale of atomic dimensions and smaller, it is simply not possible to describe phenomena with the completeness expected from classical physics. Some of the measurements or knowledge required for a complete description from the classical viewpoint, such as position and momentum, are contradictory or mutually exclusive, if the definitions of these quantities are properly understood in terms of how they are to be measured. Whenever one of the quantities is measured beyond a certain level of precision, the other is distorted to such an extent that the disturbing effect of the measurement on it cannot be determined without interfering with the original measurement. This result is inherent in the definitions of the quantities being measured. The principle goes on to state that there is cause and effect, but what quantities can be used in describing a cause-and-effect relationship have to be understood. Such things as position and momentum of an electron in an atom are not the proper quantities to measured but rather the state function (an alternative and better name for the wave function of the electron is the atom).

On another level, the idea that there are fundamental limits on the measurement of certain quantities in physics has suggested to workers in other fields that there may be analogous fundamental limits in their own disciplines on quantities to be measured or defined. The attempt to measure these quantities beyond a certain level of precision inherently distorts certain other complementary quantities. Thus the idea of complementarity may be significant for other branches of knowledge.

There is very often a strong temptation to apply major principles of physics in an overly simplistic manner to the description of the universe at all levels. Thus, following Newton's work, some argued that the universe was predetermined in all aspects and behaved like a large elaborate clockwork. This influenced the early development of the social sciences. Following the development of the entropy concept and the understanding of the second law of thermodynamics, still others asserted that the universe, taken as a whole, must be running down. Some social critics have suggested that the degradation associated with the second law is evident in modern art and even in modern society. Quantum mechanics, on the other hand, has been hailed as making possible free will and an end to all considerations of predeterminacy. Obviously, such conclusions are overdrawn.

How then should the application of physics to fields outside the physical sciences and technology be viewed? Aside from enhancing our appreciation of the beauty and grandeur of the physical universe, physics also offers new ideas and outlooks which must be justified primarily by the needs of the other areas of endeavor. Thus physics can furnish useful analogies on which to base new ways of analysis and insightful figures of speech.

Such uses of physics are sometimes apparent in literature. Indeed, concepts of modern physics, both from relativity and quantum theory, have affected many works of modern fiction. Such writers as Joseph Conrad, Lawrence Durrell, Thomas Pynchon, William Gaddis, Robert Coover, and Robert Pirsig have employed various of these concepts in their writing, with varying degrees of success. This is discussed in the book by Friedman and Donley on Einstein and in the article by Alan J. Friedman on contemporary American fiction, cited in the references for this chapter.

Murray Gell-Mann
(American Institute of Physics, Meggers Gallery of Nobel Laureates)

8 Conservation Principles and Symmetries

Fundamentally, things never change

Since at least as early as the times of the ancient Greeks, humanity has considered that there are a few fundamental building blocks, or particles, of which everything is made. Leucippus and Democritus (around 400 B.C.) believed that matter can be divided only so far, at which point the smallest possible particle will be reached. Democritus called this most fundamental particle an atom (which means "indivisible"). Aristotle realized that there must be more than one kind of fundamental particle in order to account for the different characteristics of different materials. Aristotle considered everything on Earth to be comprised of various amounts of earth, water, air, and fire, his fundamental elements.

With the development of the science of chemistry, it was recognized that there are about a hundred different elements and that all matter consists of various combinations of these elements. These elements range from hydrogen and helium, the lightest ones, to uranium, the heaviest naturally occurring element. In 1803 John Dalton, an English chemist, proposed that each element has a characteristic atom that cannot be destroyed, split, or created. He proposed that atoms of the same element are alike in size, weight, and other properties, but are different from the atoms of other elements. The name *atom* was chosen because of the clear relationship with the original concept of Democritus.

We now know that Dalton's atoms are divisible and are made up of more truly "fundamental" particles. By the 1930s the atom was known to consist of a nucleus containing neutrons and protons with electrons circulating around the nucleus (see

Chapter 7). For a time, these three particles were considered the basic building blocks of matter. However, numerous experiments performed since the 1940s indicate that neutrons and protons are made up of even more basic particles. In fact, these experiments performed at various high-energy particle accelerators (so-called atom smashers) revealed hundreds of different kinds of particles that can be produced in collisions between atomic nuclei. Throughout the 1950s and 1960s physicists became increasingly concerned about this proliferation of known subnuclear particles and wondered if they could all be "fundamental" building blocks.

Even as the number of subnuclear particles discovered experimentally was continuing to increase, many physicists were struck by the fact that many reactions expected among the various subnuclear particles were not observed. The only explanation for these missing reactions seemed to be that there were new kinds of conservation laws which would apparently be violated if these reactions took place. Slowly these new conservation laws were discovered. From previous experience, physicists knew that each conservation law implied a certain structure or symmetry in nature.

In 1961, Murray Gell-Mann at the California Institute of Technology and Yuval Ne'eman, an Israeli physicist, discovered an important new classification scheme of the subnuclear particles based on the symmetries implied by the new conservation laws. This new classification scheme, in turn, led Gell-Mann to suggest that the large number of subnuclear particles were all made up of only a few particles, which he named quarks. Subsequent research has led physicists to accept the quark model as correct, and physicists hope that the quarks are now indeed the long-sought fundamental building blocks of nature.

This chapter begins by discussing the structure of the nucleus, starting with our understanding about the time of the discovery of the neutron in 1932. We will follow the subsequent evolution of the understanding of the nature of the nucleus and the amazing force that holds it together and discuss the discovery of some of the new subnuclear particles. Following a brief summary of our knowledge of subnuclear particles as of about 1960, we will see how the careful study of conservation laws governing the interactions among the subnuclear particles led to the discovery of important new symmetries in the physical laws governing characteristics of subnuclear particles. We will see how these new symmetries led to the quark model and study its problems and successes. Finally, we will try to summarize our present knowledge of the "fundamental" building blocks and indicate possible future developments. The search for the fundamental building blocks of nature has a long scientific history. One of the main goals of this chapter will be to see how the study of conservation laws and their related symmetries has led to one of the most significant developments in the history of physics.

The Nuclear Force and Nuclear Structure

As discussed in the preceding chapter, we know that the atom is a nuclear atom; that is, it has a nucleus. The nucleus contains more than 99.9 percent of the mass of an atom. Furthermore, this mass is confined to a very small volume. Various experiments indicate that the diameter of a typical nucleus (e.g., a carbon atom nucleus) is less than 10^{-14} meters (one hundred-thousandth of a billionth of a meter), which is only about 0.01 percent of the diameter of the atom. Because nearly all the mass of an atom is concentrated into such a very small volume, the density of matter in a nucleus is very high. The density of matter in a nucleus is about 10^{17} kg/m^3 or a hundred million billion kilograms per cubic meter!* We should not be surprised to find that there are new physical phenomena involved with the structure of the nucleus that are not seen at all on a larger, more familiar, scale.

For more than fifty years it has been known that the nucleus contains protons and neutrons. The neutron mass is slightly greater than the proton mass and both are about 1.7×10^{-27} kilograms. The diameter of either a neutron or proton is known to be slightly more than 10^{-15} meters. A proton has a positive charge exactly equal in magnitude to the negative charge on an electron; a neutron has no net charge. The number of protons in a nucleus determines what element the neutral atom will be (a neutral atom has a number of electrons surrounding the nucleus exactly equal to the number of protons in the nucleus). Nuclei generally have about equal numbers of neutrons and protons, except for the heavy elements such as lead, bismuth, and uranium, which have significantly more neutrons than protons. Nuclei with the same number of protons but different numbers of neutrons are said to be isotopes of the same element. They will have the same numbers of surrounding electrons and thus will behave identically chemically.

It is important to note that if we consider only the two basic forces we have discussed so far, we must conclude that any nucleus would be completely unstable. A nucleus comprised of protons and neutrons is a nucleus containing many "like" charges and no "unlike" charges. Because like charges repel, we know that the electromagnetic force is trying to break all nuclei apart (except for hydrogen, which has only one proton in its nucleus). What then, holds nuclei together?

The only other force discussed so far is the gravitational force. If we try to see whether the gravitational attraction between the neutrons and protons in a nucleus can hold the nucleus together, we find that gravity is much too weak. If we assume

*Numbers such as 10^{-14} or 10^{17} are referred to as numbers in "scientific notation," where the exponent indicates how many decimal places to move the decimal point. Thus 10^{17} is the number 1 followed by 17 zeros and the decimal point, and 10^{-14} is the number 1 preceded by 13 zeros and the decimal point.

that the gravitational force between neutrons and protons in a nucleus is given by the same formula that correctly yields the gravitational attraction between the Moon and the Earth, or the Earth and the Sun, or any two large objects, we find that gravity is weaker than the electromagnetic repulsion by a factor of about 10^{39}!

Clearly, another kind of attractive force exists in order to explain why nuclei are stable. This new force is called the strong nuclear force and has been studied for over fifty years. It acts between neutrons and protons in a nucleus but has no effect on the surrounding electrons. It is about a hundred times stronger than the electromagnetic repulsion. Thus nuclei with more than about a hundred protons are unstable because the electromagnetic repulsion, which continues to add up, is finally greater than the attractive nuclear force, which does not continue to add up (physicists say it saturates).

The strong nuclear force is very different from either the electromagnetic or gravitational force. Both of the latter are infinite-range forces (their forces act at all distances), although they become weaker at larger distances (the inverse square law discussed earlier). In contrast, the strong nuclear force is finite-ranged. Two nucleons (either neutrons or protons) must almost, but not quite, touch each other before the strongly attractive nuclear force "turns on." This finite-ranged characteristic of the nuclear force has resulted in it sometimes being referred to as the nuclear "glue," because glue works only between objects brought into contact with each other. The strong nuclear force is different from the electromagnetic and gravitational forces also in that there is no known simple mathematical expression that permits one to calculate the strength of the force for a given situation. The nuclear force is known to depend on several different variables in rather complicated ways. Although nuclear physicists have studied the nuclear force for many years, they have been unable to find any simple expression to describe the strength of the force. Later in this chapter we will see why this situation exists.

In addition to the electromagnetic, gravitational, and strong nuclear forces, one more basic force is known to exist. This force is known as the weak nuclear force and is responsible for a certain kind of radioactive decay of nuclei known as beta decay. Radioactive decay refers to the spontaneous emission by the nucleus of one or more particles. The decaying nucleus transmutes itself into another kind of nucleus. In the simplest form of beta decay, a neutron inside a nucleus changes into a proton and an electron, and the electron is emitted from the nucleus. The weak nuclear force is a factor of about 10^{11} times weaker than the strong nuclear force, and we need not consider it further for our simple discussion of nuclear structure. There are several other kinds of radioactive decay besides beta decay, in which other kinds of particles may be emitted.

Historically, studies of naturally radioactive substances provided our first clues regarding nuclear structure. The most commonly emitted particles from radioactive elements were given the names of the first three letters of the Greek alphabet—they were named alpha, beta, and gamma particles. We now know that an alpha particle is the nucleus of a helium atom, which consists of two protons plus two neutrons; that a beta particle is an electron; and that a gamma particle (more often called a gamma ray) is a high-energy quantum of electromagnetic energy, also known as a photon. Although nuclear physicists still study natural radioactive decay, more often they study nuclei by bombarding stationary target nuclei with high-velocity beams of various particles, such as protons, alpha particles, or electrons. Such collisions provide a way to add particles and energy (from the kinetic energy of the incident particle) to nuclei in a precisely determined way, in order to study how nuclei transmute into other nuclei and break up into constituent particles. Such experiments are called nuclear reactions because they result in the separation or production or combination of the basic particles, or groups of particles, from the nucleus. Nearly all of the information presented here regarding nuclear structure and the kinds of particles found inside nuclei has been obtained from careful studies, performed at various accelerator laboratories around the world, of particle collisions with nuclei.

Shortly after the basic characteristics of the strong nuclear force were realized (namely, its strength and finite range), an extremely important step was taken toward understanding the basic nature of this force. In 1935, the Japanese physicist Hideki Yukawa proposed a theory that explained the nuclear force as due to an exchange of finite-mass quanta (or particles). This kind of description of a force is known as a quantum-field theory because the force field is suggested to consist of "virtual" quanta that are exchanged between nucleons when they come within the range of the nuclear force. A "virtual" particle is one that may exist for only a very short time and cannot be observed experimentally except by adding a large amount of energy to make it "real." Virtual particles are allowed by the Heisenberg uncertainty principle.

A quantum-field theory was known to exist for the electromagnetic force and was first suggested for the strong nuclear force by Werner Heisenberg in 1932. In the quantum-field theory of the electromagnetic force, the quanta that are believed to be exchanged between charged particles are the quanta of light, or photons. Each electrical charge is surrounded by a cloud of virtual photons, which are being constantly emitted and absorbed by the charged particle. When two charged particles are brought near each other they exchange virtual photons, so that one charged particle emits a photon and the other charged particle absorbs it, and vice versa. This constant interchange between the two results in the electrical force between them. In this quantum-mechanical theory of force fields, it is the zero mass of the photons that makes the electromagnetic force infinite-ranged.

According to Yukawa, because of the short range of nuclear forces, a nucleon is surrounded by a cloud of virtual quanta with mass, which the nucleon is constantly emitting and absorbing. If another nucleon is brought close to the first, a particle emitted by one may be absorbed by the other, and vice versa. This exchange of particles results in a force being exerted between the two nucleons. Yukawa was able (using Heisenberg's uncertainty principle) to estimate the mass of the exchange particles. The result was that, if his quantum-field theory was correct there should exist a new particle with mass about one-seventh of the neutron or proton mass. Note that this mass is still about 270 times the mass of an electron, so that Yukawa's new particle clearly did not correspond to any known particle at that time. In contrast to the quantum-field theory for the electromagnetic force, Yukawa's proposal required that there be a new, undiscovered particle. Clearly, the experimental discovery of Yukawa's particle would provide strong support for the quantum-field description of forces.

In 1947, the predicted particle was discovered to be emitted in nuclear reactions induced by high-energy cosmic rays (mostly protons) from space. The particle was named the pi-meson or, for short, the pion. In 1949 Yukawa was awarded the Nobel Prize in physics for his work on this subject.

Subsequent theoretical and experimental investigations showed that the nuclear force is not explained simply by the exchange only of pi-mesons. It is now known that to account for various characteristics of the nuclear force within the framework of the quantum-field theory, one must include the simultaneous exchange of more than one pi-meson and even the exchange of other, more massive, field quanta. Nevertheless, the basic correctness of Yukawa's theory is almost universally accepted.

In fact, it is now believed that all basic forces are properly described by quantum-field type theories. The gravitational force, because it is infinite-ranged (like the electromagnetic force), must have a zero-mass exchange quantum. The necessary particle has been named the graviton and is generally believed to exist, even though it has not been observed experimentally. Because the gravitational force is so weak, each graviton is expected to carry a very minute amount of energy. Such low-energy quanta are expected to be difficult to detect and most physicists are not surprised that the graviton has escaped experimental verification. Several groups of scientists around the world are performing elaborate (and difficult) experiments to try to detect gravitons.

On the other hand, the exchange particles for the weak nuclear force, called the intermediate-vector bosons, are so massive that no present particle accelerator (atom-smasher) can deliver enough kinetic energy to "make" these particles in the usual way by bombarding stationary targets with high-energy beams of particles. The weak nuclear force is not only very much weaker than the strong nuclear force but also very much shorter-ranged. Thus by the same kind of argument that enabled Yukawa to estimate the mass of the pi-mesons, nuclear physicists knew that the masses of the intermediate-vector bosons must be very large. In 1982 and 1983, physicists at the European Center for Nuclear Research (CERN) used a new experimental technique involving colliding beams of high-energy protons that enabled them to achieve much higher reaction energies than would be possible with stationary targets. They were able to identify the intermediate-vector bosons in the reaction products. The masses of these particles were found to be close to those previously estimated and are about a hundred times the mass of a proton.

It is interesting to note that the concept of quantum-field theory, which "explains" a force as the exchange of virtual particles, is actually just the latest attempt by scientists to understand the "action-at-a-distance" problem discussed earlier in this book. Just how two objects, not directly in contact with each other, can have effects on each other has long been a serious, fundamental question (raised even in Newton's time). The hypothetical ether, primarily postulated to explain how light travels through empty space, was also believed to help explain the action-at-a-distance problem. Now, the quantum-field theory, generally accepted by most physicists, may finally have provided the explanation of this long-considered problem. Even though Yukawa's original idea for the description of the strong nuclear force as due to the exchange of only pi-mesons is known to be incorrect, his prediction of a new particle based on his theory, and its experimental discovery, led to the general acceptance of the quantum-field description of the basic forces. Later in this chapter we will see that the reason why more than one kind of exchange particle is required to describe the strong nuclear force may be because the strong nuclear force is not actually one of the basic forces; rather, it is more correctly described as a residual interaction left over from a much stronger, more basic force that exists inside nucleons.

Yukawa's theory had another important effect on the development of nuclear physics. His prediction of yet another particle besides neutrons and protons that might exist inside nuclei provided much of the original motivation to search for new kinds of "fundamental" particles. In fact, a different particle, now called the muon, was discovered shortly before the pion, and was, for a short time, mistakenly thought to be Yukawa's predicted exchange particle. We now know that the muon is not at all like a pion, but is basically like an electron except that it is about two hundred times heavier. The discoveries of the muon and the pion were just the beginning. With the aid of new, more powerful accelerators, more and more kinds of particles were discovered. By about 1957, eighteen so-called elementary particles were known, including ones with names such as sigma, lambda, xi, and K-meson (see Table 8-1). During the late 1950s and 1960s, the list grew until only numbers could be assigned for names of the new particles and well over a hundred "elementary" particles were known to exist. Such a proliferation of particles removed the apparently simple description of the basic building blocks of matter known in the 1930s. Physicists generally were dismayed by the apparently complicated collection of fundamental particles. The Greek idea that simple is beautiful continues to dominate scientific thinking.

Even before the number of known subnuclear particles began to grow so rapidly, nuclear physicists recognized that the different particles fell naturally into a few different groups or "families" according to their characteristics. For each particle, physicists tried to determine mass, electric charge, spin (a quantum number describing its intrinsic angular momentum), and some other properties to be discussed below. The particles were then assumed to be in families with other

Table 8-1 Fundamental Particles (circa 1957)

Family	Particle	Symbol	Charge	Spin	Mass (MeV)	Lifetime (sec)
Photon	photon	γ	0	1	0	Stable
Lepton	electron	e^-	−1	½	0.511	Stable
	muon	μ^-	−1	½	105.7	2.2×10^{-6}
	electron neutrino	ν_e	0	½	0?	Stable
	muon neutrino	ν_μ	0	½	0?	Stable
Meson	charged pion	π^+	1	0	139.6	2.6×10^{-8}
	neutral pion	π^0	0	0	135.0	0.8×10^{-16}
	charged Kaon	K^+	1	0	493.8	1.2×10^{-8}
	neutral Kaon	K^0	0	0	497.7	0.9×10^{-10} or 5.2×10^{-8}
	eta	η	0	0	548.7	$\sim 2 \times 10^{-19}$
Baryons	proton	p	1	½	938.3	Stable
	neutron	n	0	½	939.6	917
	lambda	Λ	0	½	1115.4	2.6×10^{-10}
	sigma plus	Σ^+	1	½	1189.4	7.9×10^{-11}
	neutral sigma	Σ^0	0	½	1192.3	5.8×10^{-20}
	sigma minus	Σ^-	−1	½	1197.2	1.5×10^{-10}
	neutral xi	Ξ^0	0	½	1314.3	2.9×10^{-10}
	xi minus	Ξ^-	−1	½	1320.8	1.6×10^{-10}

particles possessing similar properties. As an example, the eighteen "elementary" particles known in 1957 are listed in Table 8-1 separated into the four known families.

The baryons are particles with rest (that is, proper) masses equal to or greater than the rest mass of the proton. (*Baryon* means "heavy particle.") These particles are all known to interact with each other via the strong nuclear force (as well as the electromagnetic force if they are charged). Note that they all have half-integer spin as one of their inherent properties. Except for the proton, all the baryons are known to be unstable and usually will spontaneously decay (break up into other particles) in a small fraction of a second. A short lifetime refers to the fact that the particle will exist for only a short time after it is created in a nuclear reaction. When it decays, it turns into one or more longer-lived particles that will in turn decay until only stable particles (like protons and neutrons) are left. (Even a neutron, when isolated, will spontaneously decay with a mean lifetime of about fifteen minutes, and physicists are currently trying to determine if a proton actually is completely stable—that is, has an infinitely long lifetime.)

The mesons are particles with rest masses less than the baryons but greater than the leptons, the next family. These particles all have zero or integer intrinsic spin values. Besides a neutral and a charged pi-meson (Yukawa's predicted particle), there exists a neutral and a charged K-meson, or kaon. Note that the mesons are all unstable, with a characteristic lifetime of a small fraction of a second.

The leptons have rest masses less than those of the mesons and have half-integer spin. (*Lepton* means "light particle" and *meson* means "medium particle.") This family consists of the electron and the muon, which acts like just a heavy electron. The neutrinos are each associated with either an electron or a muon and either have zero rest mass or at most a very small rest mass close to zero.

The photon is the quantum of electromagnetic radiation and is in a family all by itself. It has spin equal to one and (probably) zero rest mass.

Besides the particles listed in Table 8-1, for each particle there was known (or believed) to exist a corresponding antiparticle. An antiparticle has the same rest mass, intrinsic spin, and lifetime as its associated particle. It has, however, an exactly opposite sign of electric charge and certain other characteristics not yet discussed. The antiparticle of an electron, for example, is called a positron; it was discovered in 1932. The antiparticle of a proton is called simply an antiproton and was first observed experimentally in 1955. Some particles such as the neutral pion (t°) and the photon are believed to be their own antiparticles. When a particle and its antiparticle collide, they will annihilate each other and produce new particles and/or electromagnetic energy (photons). If we count antiparticles separately, then the list of subnuclear particles in Table 8-1 actually indicates thirty-two particles (fifteen particles and their antiparticles, plus the t° and photon).

The list of particles in Table 8-1 might not be considered too long and complicated, expecially when organized into families as indicated. Unfortunately, the list began to grow rapidly in the late 1950s and throughout the 1960s. The proliferation of new subnuclear particles arose in the family identified as the baryons. More and more particles were discovered with larger rest masses and shorter decay times. Their lifetimes became so short that these new "particles" began to be called simply resonances, which directly implied that physicists were no longer sure that these new energy quanta were actually particles. Elementary-particle physicists gave up trying to find names for all these new baryons and labeled them simply by their rest mass energies. The somewhat complicated situation of Table 8.1 became much more complex and distinctly bewildering. The remainder of this chapter will be devoted to a discussion of how this proliferation of new particles was finally understood in terms of a simpler underlying structure by the study of conservation laws and their related symmetries.

Conservation Laws and Invariants

Physicists have recognized the importance of conservation laws for some three centuries. Newton presented his laws of conservation of mass and momentum in his *Principia* in 1687. The development of the law of conservation of energy was one of the most important nineteenth-century advances in physics (Chapter 4). Nuclear physicists studying nuclear reactions in carefully controlled experiments soon realized that several different quantities were always conserved. Some of these were the already-known conserved quantities such as energy, momentum, electric charge, and angular momentum. But eventually new quantities were discovered to be conserved, and physicists suspected that the careful study of these quantities would lead to a better understanding of the mysterious nuclear force.

Recall that conservation of a quantity simply means that it remains constant in total amount. Nuclear physicists can carefully prepare a nuclear reaction so that they know exactly how much total mass, energy, electric charge, momentum, and so on, is available before the reaction takes place. By the use of specialized instruments, they can also measure these same quantities after the reaction takes place. A careful comparison of the known amounts of each quantity before and after the nuclear reaction reveals which quantities are conserved (do not change). In such studies it was discovered that nuclear reactions controlled by the strong nuclear force obey more conservation laws than are obeyed by gravitational and electromagnetic interactions.

In any interaction (whether between two billiard balls or between two subnuclear particles), the following seven quantities are always conserved:

1. Mass-energy
2. Momentum
3. Angular momentum
4. Electric charge
5. Electron family number
6. Muon family number
7. Baryon family number

In considering the last three quantities, one counts a particle as +1 and an antiparticle as −1. Thus these three laws say that the total excess of particles over antiparticles (or vice versa) within each family is always exactly maintained and further indicates that these "families" are somehow natural classifications. The first four conserved quantities have been known for more than eighty years (some much longer) and have been discussed earlier in this volume. These seven basic quantities are conserved by all four of the basic forces of nature, so far as we know. Let us now consider each of the basic forces separately with regard to known conservation laws obeyed by that force.

In any reaction in which the strong nuclear force dominates (and such is the case for most nuclear reactions), besides the seven basic quantities listed above, the quantities known as parity, isotopic spin, and strangeness are also conserved. The parity of a system involves its helicity or inherent right- or left-"handedness." For example, a right-handed screw would have right-handed helicity and a left-handed screw would have left-handed helicity. For nuclear particles, helicity deals with the orientation of the intrinsic spin vector of a particle with respect to its velocity vector (direction of travel). Isotopic spin is a quantum-mechanical quantity that describes the neutron or proton excess of a nuclear system. Strangeness is a quantum-mechanical quantity with no simple description and will be discussed further below. Related to these conserved quantities is the fact that nuclear reactions are also observed to obey an "operation" known as charge conjugation. This last statement means simply that for any possible nuclear reaction there is another possible reaction, which corresponds to changing all positive charges to negative ones and all negative charges to positive ones (actually, interchanging all particles with their antiparticles). Some of these new quantities, such as isotopic spin and strangeness, would not be noteworthy if they were not conserved in nuclear reactions. Because they are conserved we must regard them as clues to the fundamental characteristics of subnuclear particles.

The electromagnetic force is second in strength to the strong nuclear force, and interactions dominated by the electromagnetic force apparently conserve all the same quantities as the strong nuclear force, except for isotopic spin. That isotopic spin is not conserved by the electromagnetic force is related to the fact that isotopic spin involves neutron or proton excess—that is, the excess of particles with or without charge. Because the electromagnetic force depends primarily on whether or not an object is charged, it is sensitive not so much to the neutron or proton excess, but rather just to the number of charged particles.

The weak nuclear force obeys only the seven basic conservation laws. When it was shown in 1957 that this force did not conserve parity, physicists were generally surprised, because such a failure indicated that the universe is fundamentally not ambidextrous. Such an inherent "handedness" seemed peculiar, and the weak nuclear force was considered somewhat enigmatic.

Finally, the gravitational force is so weak that there are no known reactions between individual subnuclear particles dominated by this force. Consequently it is not known which of the conservation laws are obeyed by the gravitational force at the microscopic level.

Before we end this brief discussion of known conservation laws, it is important to remember the implications of conservation laws. Once a law of nature is recognized, we have discovered a fact of nature. A natural law (if it is correct) must be obeyed; there is no choice. A civil law which says that one must obey stop signs can be ignored (perhaps with some unpleasant consequences). A natural law is a constraint on how physical systems can develop. For example, when we know that baryon number is conserved, we have automatically ruled out a large number of nuclear reactions which would not conserve this quantity. By the time we add the constraints of conserving the other family members, charge, parity, and so on, we can often predict that only a few nuclear reactions are possible for a given set of starting conditions. If any of the other reactions are observed experimentally, one of our assumed conservation laws is wrong.

The history of the study of subnuclear particles actually proceeded in just the opposite way. Although more and more subnuclear particles were discovered, most of the possible reactions expected between these particles did not occur. In order to explain why these reactions were forbidden, new conservation laws had to be invented. As it turned out, some of these newly discovered quantities, such as isotopic spin and strangeness, eventually led to a new understanding of the fundamental building blocks of nature.

Conservation Laws and Symmetries

The new conservation laws led to a new understanding of the fundamental building blocks of nature because of symmetries that these conservation laws implied. Symmetries in nature are something with which we are all familiar. Many natural objects have a left side which is just like the right side, or a top that is identical to the bottom. Many flowers, crystals, snowflakes, and even human faces are either perfectly symmetric or almost symmetric. By symmetric, in these instances, we mean there is some rotation that can be applied to the object that will result in it looking the same (or nearly the same) as when it started. (This interest in symmetry was at the root of the ancient Greek search for perfection in the heavens as discussed in chapter 2.) For our usual sense of symmetry we mean mirror symmetry, which interchanges the right and left sides. A snowflake, however, might be symmetric through several different rotations, such as 60, 120, and 180 degrees, and so on. Figure 8-1 shows a highly symmetric snowflake that will look exactly the same when rotated through any angle that is a multiple of 60 degrees.

Physicists have generalized our common understanding of symmetry to say that a symmetry operation is any well-defined action which leaves the object looking the same as when it started. Physical symmetry operations must be expressed as mathematical operations. For example, the symmetry operations discussed for a snowflake are rotations. Of interest for our subject here is the realization that the basic laws of physics are left unchanged by many of these so-

Figure 8-1. Snowflake with 60° symmetry (Richard Holt/Photo Researchers)

called symmetry operations. For example, none of the laws of physics is changed in a closed system if the entire system is rotated by any arbitrary angle. Similarly, the laws of physics are unchanged if a closed system is simply moved in space (provided that all relevant objects are moved with it) or if the system is displaced in time. These are three examples of symmetry operations which can be performed.

The three symmetry operations mentioned above which do not change the laws of physics appear trivial, and to some extent they are. We would not expect scientists in China to discover that their laws of physics are different from ours or scientists a hundred years from now to find different physical laws governing the universe. However, each of these trivial symmetry operations corresponds to one of the known conservation laws. Using simple mathematical analysis, it can be shown that the fact that the laws of physics are unchanged by a translation (change of location) in space corresponds to the law of conservation of momentum. *Corresponds* means that the one necessarily implies the other. The fact that the laws of physics are unchanged by a translation in time corresponds to the law of conservation of energy. That the laws are unchanged by a rotation corresponds to the law of conservation of angular momentum.

The relationships between symmetry operations and conservation laws do not end with just these three "trivial" symmetry operations. The combination of all three (space, time, and rotation) corresponds to Einstein's relativity principle, sometimes known as Lorentz invariance. The fact that the laws of physics are unchanged by a change of phase in the quantum-mechanical wave function (see

Chapter 7) corresponds to the law of conservation of electric charge. The symmetry operations that correspond to the other known conservation laws are known but are mainly operations in the mathematical formulation of quantum mechanics.

The correspondence between a symmetry operation and a conservation law works both ways; either one implies the other. This correspondence makes these relationships extremely important. Every new conservation law implies a new "quantum number" and a symmetry operation that will leave the laws of physics unchanged. From every newly discovered conserved quantity a new symmetry in the laws of physics may be inferred. It was precisely such a relationship that enabled Gell-Mann and Ne'eman to discover the new symmetry implied by the known conservation laws for nuclear reactions, including the then newly discovered conserved quantity called strangeness.

The Quark Model

Now that we have discussed the important relationship between conservation laws and underlying symmetries, we can consider how the discovery of new conserved quantities led to the development of a new model for the fundamental building blocks of nature. This new model started when, independently, Gell-Mann and Ne'eman recognized that if baryons and mesons (two of the families of particles discussed above) were organized by their strangeness and isotopic spin quantum numbers, simple patterns emerged. Recall that strangeness is a quantum number introduced to explain why many expected reactions among the known particles were observed relatively infrequently and that isotopic spin described the different charge states that strongly interacting particles can have.

These simple patterns can be seen most easily if we use a new quantity Y, called hypercharge defined as

$$Y = S + B$$

where S is the strangeness quantum number and B is the baryon number. The simple patterns obtained for baryons with intrinsic spin of 1/2 and for mesons with intrinsic spin of 0 are shown in Figure 8-2. The location of each particle is determined by the values of its hypercharge quantum number Y and its isotopic spin quantum I_3 (actually the so-called z-component of the isotopic spin quantum number). The location of each particle is uniquely determined. The symbols Σ, Λ, and so on, refer to the particles listed in Table 8-1. Note that the baryon classification includes the well-known neutron and proton plus some of the newer, "stranger" particles.

The meson classification includes the pi-mesons originally predicted by Yukawa plus some recently discovered mesons. Both of these patterns are hexagons with a total of eight members of each (two at the center). Gell-Mann referred to these classifications as the eight-fold way. The important concept here

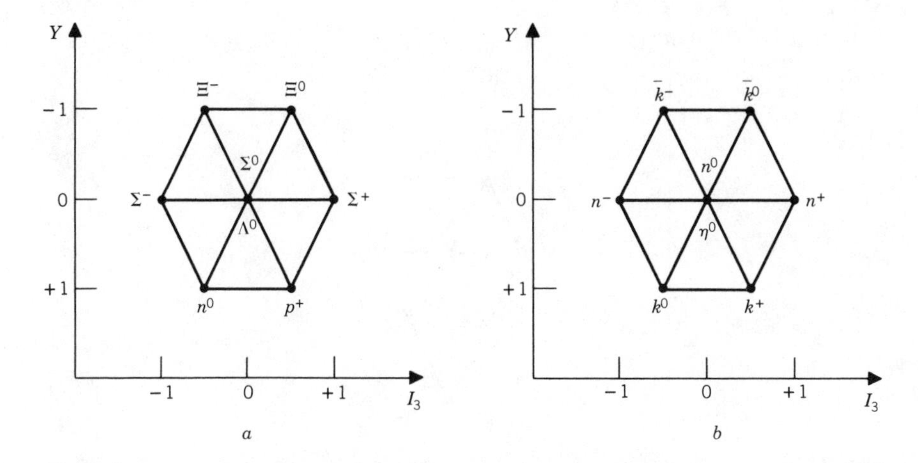

Figure 8-2. Classification diagrams. (a) Spin 1/2 baryons. (b) Spin 0 mesons. *Y*, hypercharge; I_3, z-component of isotopic spin.

is not what hypercharge and isotopic spin are, but that by using them simple patterns are obtained. It is equally important to remember that these quantum numbers were "invented" (or discovered) in order to obtain needed conservation laws that could explain why some nuclear reactions were allowed and others were forbidden.

In physics, it is often said that the real test of any new model or theory is whether it can correctly predict something new. Although Gell-Mann's and Ne'eman's new way of classifying particles clearly leads to some simple patterns, unless their classifications lead to some new knowledge not previously known it would all be merely an interesting scheme. Remarkably, the new classification scheme immediately provided a prediction of a new particle that had never been observed experimentally. Figure 8-3 shows the plot for baryons with intrinsic spin of 3/2 similar to the plots of Figure 8-2. All of the particles represented on the plot were known except for the particle called Ω^- (omega minus) shown at the very bottom of the figure. The pattern (or symmetry) of the figure is obvious: it is an upside-down triangle. Simply by its location in the figure, Gell-Mann could predict the hypercharge (and therefore the strangeness and the isotopic spin of the new particle). Further considerations allowed him to determine the charge and to estimate the mass of the Ω^- as well.

Given the predicted properties of this new particle, experimental physicists at the Brookhaven National Laboratory on Long Island, New York, immediately set out to discover if this new particle, the Ω^-, actually existed. The general plan was to try to pick out for study at least one reaction which should lead to the production of the Ω^-, while still obeying all known conservation laws. If the Ω^- was created, then the experimenters would be able to recognize it by its mass, charge, spin, and

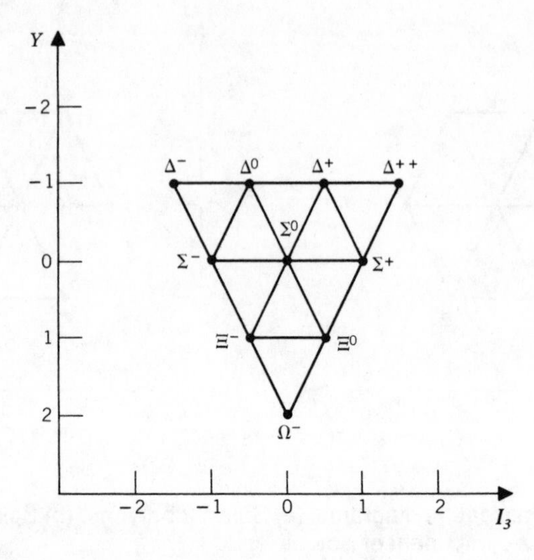

Figure 8-3. Baryon decuplet classification diagram. Y, hypercharge, I_3, z-component of isotopic spin.

other quantum characteristics. A likely reaction was selected and, in a dramatic verification of the fundamental importance of the new classification scheme, the Ω^- particle was observed experimentally in November 1964, only seven months after the prediction of its existence.

The experimental discovery of the Ω^- particle clearly demonstrated that there was a basic significance to the new classification scheme. In 1963, Gell-Mann showed that the groupings of the particles according to their patterns in the plots of hypercharge versus isotopic spin were accurately described by a branch of mathematics called group theory (not surprising, because the particles were separated into groups). In particular, these different groups were described by a mathematical representation known as $SU(3)$. Again, it is not important that we explain the mathematics of $SU(3)$ here except to know that it describes groups that comprise combinations of only three fundamentally different objects.

Thus Gell-Mann (and another physicist, George Zweig) hypothesized that the known baryons and mesons (the families described by the new classification) were comprised of three even more fundamental particles. Gell-Mann named these new particles quarks (from a simple rhyme in James Joyce's novel *Finnegan's Wake*). Because the great proliferation of new subnuclear particles was occurring in the baryon and meson families, Gell-Mann's hypothesis, if successful, would result in a great simplification in the list of fundamental particles.

Gell-Mann then deduced the characteristics such as mass, electric charge, spin, and so on that each of the three quarks must have for combinations of them to describe the known characteristics of the baryons and mesons. Some of the characteristics of the quarks seemed peculiar, especially the need for them to have charges that are either one third or two thirds of the charge of an electron or proton. It had been thought for many years that the charge on an electron was the smallest unit of electrical charge. Gell-Mann found that baryons must each be made up of three quarks and mesons of two quarks. Corresponding to each of the three quarks is an antiquark that has the same relationship to its quark as other antiparticles to their corresponding particles (such as opposite charge, same mass, and the like, discussed earlier). A meson is actually a quark and antiquark pair. All of the known baryon and meson groups (in the new classification scheme) were accounted for on the basis of the quark model. Even the existence of the Ω^- particle was correctly predicted by the quark model.

The apparent success of, and consequent simplification resulting from, the quark model was great enough that many experimenters immediately set out to try to detect the quarks experimentally. Many physicists were skeptical from the start as to the likelihood of success, because no particles with a fractional charge had ever been seen and each of the three quarks is predicted to have a fractional charge (either one third or two thirds of the electron charge). To date, no convincing experiments have been able to detect an individual quark, and no free quark has been observed experimentally. To further complicate the situation, recent discoveries of more massive subnuclear particles have shown that the original three-quark model must be extended to include a fourth, fifth, and sixth quark. The quark model will not represent such a simplification if too many are required.

In spite of the failure to detect an individual quark and the recent increase in the number of needed quarks, most nuclear physicists today are convinced that the quark model is basically correct. In fact, the need to add new quarks has provided some of the strongest evidence for the validity of the model. When a new particle was discovered which required a new quark in order to have its characteristics properly described, one could immediately predict other new particles that would correspond to other combinations of the new quark with the "old" quarks. These new particles have consistently been discovered experimentally, further verifying the existence of the new quark and the correctness of the entire quark model.

The properties of the six known quarks are listed in Table 8-2. Gell-Mann's original three quarks are usually named *up*, *down*, and *strange*. They can account for all of the baryons and mesons of the thirty-two "original" elementary particles of Table 8-1. A proton, for example, is believed to consist of two up quarks and one down quark; a neutron of one up and two down quarks; and Yukawa's positively charged pi-meson of an up quark and an antidown quark.

The discovery of a new, relatively long-lived particle known as the J or ψ (psi) particle in 1974 required the introduction of a fourth quark, usually called the *charmed* quark. The discovery of another relatively long-lived particle called the γ (upsilon) particle in 1977 required the fifth quark called the bottom (or beauty) quark. Some experiments performed in 1983 indicate the existence of a sixth quark, called the top (or truth) quark. Besides these six kinds (or flavors, as they

Table 8-2 Quark Characteristics

Name	**Spin**	**Electric Charge**	**Baryon Number**
up	1/2	+2/3	1/3
down	1/2	− 1/3	1/3
charmed	1/2	+2/3	1/3
strange	1/2	− 1/3	1/3
top (truth)	1/2	+2/3	1/3
bottom (beauty)	1/2	− 1/3	1/3

are often called) of quark, each quark can come in three different colors, usually chosen as red, yellow, and blue. Also, for every quark there is an antiquark, as discussed above: thus there are thirty-six different quarks, but only six basic kinds. (The use of such words as *strangeness, charm, color, up, down,* and so on to describe physical quantities, while perhaps appearing whimsical, is just as legitimate as coining new Greek words such as *entropy, enthalpy*, and so on.)

If the quarks are actually the fundamental building blocks of the baryons and mesons, there must be a new force, not yet mentioned, which holds quarks together inside a baryon or meson. This force must be extremely powerful in order to explain why free quarks are never observed. It was largely in attempting to explain this new force that the concept of color was introduced for quarks. The very strong force between quarks is believed to exist between quarks of different colors (or between a colored quark and its antiquark). This color force is understood to be somewhat analogous to the electric force that exists as an attractive force between charges of opposite sign. An electric force field can be understood as originating on a positive charge and terminating on a negative charge (or vice versa) as illustrated in Figure 8-4. The electrical force between them can bind them together. Similarly, the color force between quarks is believed to originate on a quark of some specific color and to terminate on two quarks of the other two colors or on one antiquark of the same color. One says that quarks feel the color force just as charges feel the electric force. The color-force fields for a baryon and a meson are illustrated in Figure 8-5.

It is known that, for ordinary light, equal mixtures of red, yellow, and blue yield no observable color (white). In analogy, a baryon made up of one red, one yellow, and one blue quark is said to have no net color. A meson made up of a quark and antiquark of the same color is said also to have no net color (that is, an antiquark has "anti"-color). An amazing fact is that every particle so far discovered that is understood to be made up of a combination of quarks has always turned out to be a "colorless" combination. One always finds combinations of

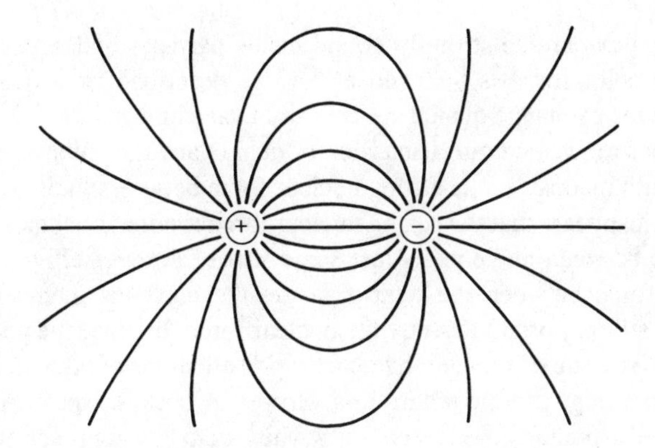

Figure 8-4. Lines of electric force field between two charges.

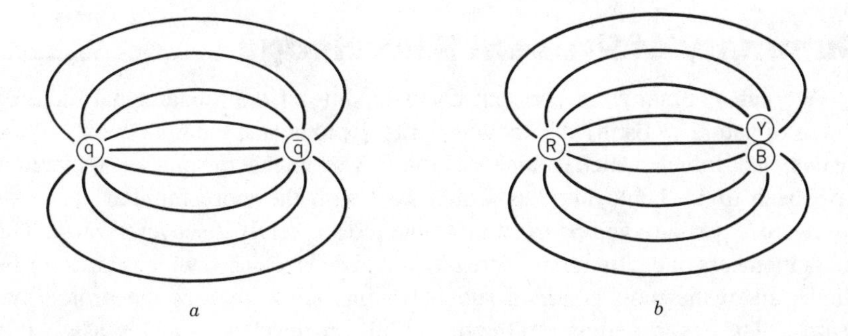

Figure 8-5. Lines of color force field. (a) Between a quark–antiquark pair in a meson. (b) Between three quarks in a baryon. *q*, quark; $\overline{q}$ antiquark; *R*, red quark; *Y*, yellow quark; *B*, blue quark.

three quarks having one of each color, or a quark-antiquark pair. It appears that quarks always come in threes or quark-antiquark pairs just so that there will be no net color. The color force requires this so that the color force can always terminate within each particle.

The color force is very strong, because quarks are apparently so tightly bound inside particles that it has not yet been possible to find a "free" quark. In addition, it is believed that the color force between quarks increases rather than becomes weaker as quarks are separated. By contrast, the electric and gravitational forces between two particles become weaker as the particles are separated. The color force thus is a rather peculiar force compared with the forces with which we are more familiar. The fact that the color force increases as quarks are separated may explain why quarks are so strongly bound inside particles and never seen individually. The color force is believed also to be described by a quantum-field theory, where the exchange quanta are referred to as gluons.

Finally, before we end our discussion of quarks and the color force between them, we should reconsider the strong nuclear force between nucleons (neutrons and protons). It appears that the strong nuclear force is not a fundamental force at all. The force between nucleons is just some sort of leftover effect of the very strong color force between the quarks inside the nucleons themselves. Nevertheless, it is still important to study the nuclear force, because the nuclear force determines the structure of nuclei that exist inside all atoms of all materials and is responsible for energy production in stars. However, because the nuclear force is actually derived from the color force, only when we try to approach the study of the nuclear force from that perspective can we hope to be completely successful in understanding it. This probably explains why nuclear physicists have been unable to obtain a simple description of the nuclear force even after fifty years of study.

Summary of Present Knowledge

We can summarize the present understanding of the fundamental building blocks of nature by listing the known quarks, leptons, and the field quanta of the fundamental forces, which is done in Table 8-3. Note that the quarks and leptons have been divided into families which start with the more familiar, or first-discovered, particles and proceed on to the most recently discovered ones. The quark members of the first family are the up and down quarks, which can describe almost all of the most common subnuclear particles, such as the proton and neutron. The lepton members of the first family are the electron and its associated neutrino. The quark members of the second family are the strange and charmed quarks, which are needed to describe the characteristics of some of the more peculiar particles in the host of subnuclear particles created in high-energy nuclear reactions. The lepton members of the second family are the muon and its associated neutrino. The muon acts very much like an electron, but it is approximately two hundred times more massive. The third family has the bottom and top (or beauty and truth) quarks, which are needed to explain the characteristics of some new particles discovered since 1977. The lepton members of the third family are the tau and its associated neutrino. The tau was discovered in 1975 and acts as an even heavier electron than the muon. Its associated neutrino has not been positively identified.

Finally, we list the field quanta of the fundamental forces. Following our discussion in the preceding section we replace the strong nuclear force with the strong color force, whose field quantum is called the gluon. Only the field quanta of the electromagnetic force and the weak nuclear force have actually been observed experimentally. The field quantum of the strong color force, the gluon, is believed to be too massive to be produced in present particle accelerators. On the other hand, because the graviton is believed to carry so little energy, it will be very difficult to observe. In spite of the fact that only two of the four field quanta of the fundamental forces have been observed, this field-quanta description of the forces is almost universally accepted as correct by modern physicists. The forces are all due to the exchange between interacting particles of particular quanta.

Note that there exists the possibility that additional families of quarks and leptons will be discovered. At this time it appears possible that the number of families might continue to grow (although certain considerations of how the universe developed in the first few moments after the big bang may set a limit on the number of families). Note also that most of the ordinary particles of the universe are made up of only the first family of leptons and quarks, and only in

Table 8-3 Fundamental Particles (circa 1986)

		Leptons		**Quarks**	
First Family	Particle	e(Electron)	ν_e(Electron Neutrino)	u(Up)	d(Down)
	Charge[a]	−1	0	⅔	−⅓
	Mass[b]	$.5 \times 10^{-3}$	0?	$\sim 5 \times 10^{-2}$	$\sim 1 \times 10^{-2}$
	Discovered	1898	1954	1963	1963
Second Family	Particle	μ(Mu)	ν_μ(Muon Neutrino)	c(Charmed)	s(Strange)
	Charge	−1	0	⅔	−⅓
	Mass	.11	0?	∼1.7	∼.3
	Discovered	1936	1962	1974	1974
Third Family	Particle	τ(Tau)	υ_τ(Tau Neutrino)	t(Truth or Top)	b(Beauty or Bottom)
	Charge	−1	0	⅔	−⅓
	Mass	1.78	0?	>30	∼5
	Discovered	1975	?	1984	1977

Field Quanta

Force	**Field Quantum**	**Mass (GeV)**	**Discovered**
Electromagnetic	Photon (γ)	0	1905
Weak	Intermediate vector boson ($W^{\pm}$, Z^0)	81,100	1983
Strong (Color)	Gluon	?	—
Gravity	Graviton	0	—

[a]Charges are in units of the charge of the proton.
[b]Masses are in units of billions of electron volts (GeV).

Fermi National Accelerator Lab

very high-energy nuclear reactions do we produce particles made up of quarks from the higher families. Thus even if more families exist, they may only be rarely involved in the universe as we know it.

Although we must remember that the leptons and quarks listed in Table 8-3 all have associated antiparticles and that each quark comes in three colors, it must be concluded that if these are the fundamental building blocks of nature the situation is fairly simple—not a hopelessly long list of seemingly unrelated particles. There is one more consideration we must discuss in order to conclude that these are truly fundamental building blocks. The question is simply "Is there any evidence that the leptons or quarks are made up of something even smaller?" Basically the question can be reduced to whether the particle has "size" or not. If a particle has spatial extent (size), then something must occupy that space. Whatever this "something" is, is then what the particle is "made of." Only if the particle is a point (that is, has no size) can it be a fundamental particle. (Note, however, that concentrating a finite amount of charge into a point would appear to require an infinite amount of energy, which has caused this question to remain somewhat controversial.)

The electron is known to be extremely small. It appears to be a pure "point" in space and has no size whatsoever. Although the limit on the size of a quark has not been set as small as for an electron, it also appears to be just a point particle. The uncertainty principle discussed in the preceding chapter can be used to help convince physicists that it is extremely unlikely that anything smaller exists inside an object known to be as small as an electron or quark. The present experimental evidence is consistent with a model of electrons and quarks as structureless particles with essentially geometric "point" properties down to dimensions as small as 10^{-16} centimeters. Thus physicists are becoming increasingly convinced that Table 8-3 is the correct list of the fundamental building blocks of nature (although there do exist some speculations regarding structure inside quarks).

What then remains to be understood? First, much work needs to be completed to verify that the leptons and quarks are fundamental. Then we need to determine exactly how many families exist. If the number of families is limited, why? Finally, physicists would like to gain a better understanding of the fundamental nature of the four basic forces. A single description can now be provided for the electromagnetic and weak nuclear force simultaneously, and some physicists feel that the strong color force will soon fall into this unified description as well. That will leave only the gravitational force to be combined into one overall unified field theory of all the forces (a goal toward which Einstein worked for more than forty years). It appears that we have made some tremendous steps toward a basic understanding of nature in the last few decades, but there remains much more to do.

In an even broader way, one can ask if it is likely that the subject of the basic building blocks of nature will end with Table 8-3. Even though there is good evidence that the quarks and leptons are indeed "fundamental" particles, and Heisenberg's uncertainty principle indicates that it is unlikely that these particles are made up of even smaller particles, one is inclined to believe that nature still has more surprises for us to discover. There have been several periods in scientific history when most scientists thought that science was nearly "complete." These periods were always followed by scientific revolutions that significantly changed our understanding of the physical universe. It seems unlikely that we have reached the end of this process. It is difficult to guess what new directions may appear next—after all, they wouldn't be really new if we already knew which way to go!

REFERENCES

This is a list of some works we think will be useful for further reading.

Chapter 1

Gillespie, Charles C. *The Edge of Objectivity*. Princeton, N.J.: Princeton University Press, 1960.

A standard book on the history and philosophy of science.

Holton, Gerald and Stephen G. Brush. *Introduction to Concepts and Theories in Physical Science*, 2nd ed. Reading, Mass.: Addison-Wesley, 1973.

A more mathematical book than this one, but with extensive discussion on the nature of scientific thought and the like and with extensive bibliographies. Chapters 3, 12, 13, and 14 are pertinent to this chapter.

Jaki, Stanley L. *The Relevance of Physics*. Chicago: University of Chicago Press, 1966.

History and philosophy of science and the effect of science on other areas of thought.

Kuhn, Thomas S. *The Structure of Scientific Revolutions,* 2nd ed. Chicago: University of Chicago Press, 1970.

This book has had a great impact on the way in which people think about "scientific revolutions."

Schneer, Cecil J. *The Evolution of Physical Science*. New York: Grove, 1964.

An interesting historical perspective of the development of physical science. Especially good discussions of the early Greek contributions.

Chapter 2

Abbott, E. A. *Flatland*. New York: Dover, 1952.

A very entertaining book, originally written a century ago, about a two-dimensional world.

Berry, Arthur. *A Short History of Astronomy: From Earliest Times Through the Nineteenth Century*. New York: Dover, 1961.

Chapter II discusses the contributions of the Greeks.

Bronowski, J. *The Ascent of Man*. Boston: Little, Brown, 1973.

Based on a television series of the same name. Chapters 5, 6, and 7 are pertinent to this chapter.

Butterfield, Herbert. *The Origins of Modern Science*, rev. ed. New York: Free Press, 1965.

A brief book, written by a historian, that discusses scientific developments up to the time of Isaac Newton.

Cornford, F. M. *The Republic of Plato*. London: Oxford University Press, 1941.

An account of Plato's Allegory of the Cave is found here.

Dreyer, J. L. E. *A History of Astronomy from Thales to Kepler*, 2nd ed. New York: Dover, 1953.

Provides a fairly detailed account of the subject matter of this chapter.

Holton, Gerald and Stephen G. Brush. *Introduction to Concepts and Theories in Physical Science*, 2nd ed. Reading, Mass.: Addison-Wesley, 1973.

An introductory text with major emphasis on the historical background of physics. Chapters 1 through 5 and pp. 154–160 are pertinent, with many references to other sources of information.

Kearney, Hugh. *Science and Change, 1500–1700*. New York: McGraw-Hill, 1971.

A small book that discusses the general scientific ferment in Europe during the period 1500 to 1700.

Koestler, Arthur. *The Watershed*. Garden City, N.Y.: Doubleday, 1960.

A slim biographical discussion of Kepler's life and work.

Koyre, Alexander. *Discovering Plato*. New York: Columbia University Press, 1945.

This book discusses Plato's views about science and philosophy.

Kuhn, Thomas S. *The Copernican Revolution*. New York: Random House, 1957.

A detailed discussion of the subject of this chapter.

Kuhn, Thomas S. *The Structure of Scientific Revolutions*, 2nd ed. Chicago: University of Chicago Press, 1970.

See comment in references for Chapter 1.

Schneer, Cecil J. *The Evolution of Physical Science*. New York: Grove, 1964.

See comment in references for Chapter 1.

Toulmin, Stephen and June Goodfield. *The Fabric of the Heavens*. New York: Harper & Row, 1961.

An interesting account of the evolution of scientific viewpoints and ideas from the time of the Babylonians to the time of Isaac Newton.

Chapter 3

Butterfield, Herbert. *The Origins of Modern Science*. New York: Free Press, 1957.

An introduction to the historical background of classical physics. Does not discuss "modern" physics.

Holton, Gerald and Stephen G. Brush. *Introduction to Concepts and Theories in Physical Science*, 2nd ed. Reading, Mass.: Addison-Wesley, 1973.

See comment in references for Chapter 2. Chapters 6 through 11 are particularly pertinent for this chapter.

Schneer, Cecil J. *The Evolution of Physical Science*. New York: Grove, 1964.

See comment in references for Chapter 1.

Chapter 4

Brown, Sanborn C. *Benjamin Thompson, Count Rumford*. Cambridge, Mass.: MIT Press, 1979.

A personal and scientific biography of one of the first American scientists.

Feynman, Richard P., Robert B. Leighton, and Matthew Sands. *The Feynman Lectures on Physics*. Reading, Mass.: Addison-Wesley, 1963, Vol. 1, Chap. 4.

A more advanced undergraduate textbook with emphasis on understanding the physical principles involved. Uses elementary calculus.

Holton, Gerald and Stephen G. Brush. *Introduction to Concepts and Theories of Physical Science*, 2nd ed. Reading, Mass.: Addison-Wesley, 1973.

Chapters 15–17, 22. See comment in reference for Chapter 2.

Mott-Smith, Morton. *The Concept of Energy Simply Explained*. New York: Dover, 1964.

A reprint of a very interesting book, originally published in 1934 and written in a descriptive style for the general public. Some of its assertions are not acceptable today.

Schneer, Cecil J. *The Evolution of Physical Science*. New York: Grove, 1964.

See comment in references for Chapter 2.

Chapter 5

Adams, Henry. *The Degradation of the Democratic Dogma*. New York: Macmillan, 1920.

This book typifies the impact of "heat death" on social thinking.

Arnhem, Rudolph. *Entropy and Art: An Essay on Disorder and Order*. Berkeley: University of California Press, 1971.

Contrasts what should be the artist's view of these subjects with the physicist's view.

Gamow, George. *Mr. Tompkins in Paperback*. New York: Cambridge University Press, 1972, Chap. 9. See comment in references for chapter 6.

Georgescu-Roegen, Nicholas. *The Entropy Law and the Economic Process*. Cambridge, Mass.: Harvard University Press, 1971.
Attempts a mathematical analysis of economic theory, employing general physics concepts from thermodynamics and quantum theory.

Gillispie, Charles C. *The Edge of Objectivity*. Princeton, N.J.: Princeton University Press, 1960, pp. 400–405.
See comment in references for Chapter 1.

Holton, Gerald, and Stephen G. Brush. *Introduction to Concepts and Theories in Physical Science*, 2nd ed. Reading, Mass.: Addison-Wesley, 1973, Chap. 18.
See comment in references for Chapter 1.

Landsberg, Peter Theodore. *Entropy and the Unity of Knowledge*. Cardiff: University of Wales Press, 1961.
Applications of entropy concept to economic theory, information theory, textual analysis.

Lewicki, Zbigniew. *The Bang and the Whimper: Apocalypse and Entropy in American Literature*. Westport, Conn.: Greenwood Press, 1984.
The latter half of this book discusses the entropy concept as employed by various American writers, including Herman Melville, Nathaniel West, Thomas Pynchon, William Gaddis, Susan Sontag, and John Updike.

Mott-Smith, Morton. *The Concept of Energy Simply Explained*. New York: Dover, 1964.
See comment in references for Chapter 4.

Powell, J. R., F. J. Salzano, Wen-Shi Yu, and J. S. Milau. "A High Efficiency Power Cycle in Which Hydrogen Is Compressed by Absorption in Metal Hydrides." *Science* 193 (23 July 1976), 314–316.
Describes a three-reservoir heat engine.

Rifkin, Jeremy. *Entropy: A New World View*. New York: Viking, 1980.
Contends that economic and energy policy must be greatly revised to take account of the second law of thermodynamics.

Sandfort, John F. *Heat Engines*. Garden City, N.Y.: Doubleday, 1962.
An introductory, essentially nonmathematical discussion of heat engines, written for the general public.

Zemansky, Mark W. *Temperatures Very Low and Very High*. Princeton, N.J.: Van Nostrand, 1964.
An introductory discussion at the college level; assumes some knowledge of physics.

Chapter 6

Bondi, Hermann. *Relativity and Common Sense*. New York: Dover, 1980.
An introductory, essentially nonmathematical discussion of relativity, written for the general public.

Born, Max. *Einstein's Theory of Relativity*. New York: Dover, 1962.
A thorough discussion, using only algebra and simple geometry, of mechanics, optics, and electrodynamics as involved in relativity theory.

Casper, Barry M. and Richard J. Noer. *Revolutions in Physics*. New York: Norton, 1972.
Chapters 12–15 provide a good low-level discussion of relativity using very simple mathematics.

Ford, Kenneth W. *Basic Physics*. New York: Wiley, 1968.
A physics textbook that uses only algebra and trigonometry, with good discussion. See Chapters 1, 4, 15, 16 and 19–22.

Gamow, G. W. *Mr. Tompkins in Paperback*. New York: Cambridge University Press, 1971.
A delightful book presenting the ideas of modern physics as distorted in the dreams of a bank clerk dozing during lectures. Chapters 1–6 deal with relativity. See footnote on p. 171 of this book for a qualification.

Hawking, S. W. "The Edge of Spacetime." *The American Scientist* 72 (July–August 1984) 355–359.
An interesting discussion of some recent ideas in general relativity, especially concerning the application of ideas of quantum theory (the subject of Chapter 7 of this book).

Holton, Gerald and Stephen G. Brush. *Introduction to Concepts and Theories in Physical Science*. Reading, Mass.: Addison-Wesley, 1973, Chap. 31.
See comment in references for Chapter 1.

Lindsay, Robert Bruce and Henry Margenau. *Foundations of Physics*. New York: Wiley, 1936.
An advanced textbook with discussions of basic metaphysical assumptions in physics. See Chapters 1, 2, 7 and 8.

Resnick, Robert. "Misconceptions About Einstein." *Journal of Chemical Education* 57 (December 1980) 854–862.
Contains interesting biographical information.

Sciama, D. W. *The Physical Foundations of General Relativity*. Garden City, N.Y.: Doubleday, 1969.
Written for the general public but contains rather sophisticated, subtle discussions.

Chapter 7

DeWitt, Bryce S. and R. Neill Graham. "Resource Letter IQM-1 on the Interpretation of Quantum Mechanics." *American Journal of Physics* 39 (July 1971) 724–738.

Feynman, Richard P., Robert B. Leighton, and Matthew Sands. *The Feynman Lectures on Physics*. Reading, Mass.: Addison-Wesley, 1965, Vol. iii.
See comment in references for Chapter 4.

Ford, Kenneth W. *Basic Physics*. New York: Wiley, 1968. Chaps. 23 and 24.
See comment in references for Chapter 6.

Gamow, G. W. *Mr. Tompkins in Paperback*. New York: Cambridge University Press, 1971. Chaps. 7, 8, 10, and 10 1/2.
See comment in references for Chapter 6.

Friedman, A.J. and Carol Donley, *Einstein as Myth and Muse*. Cambridge: Cambridge University Press, 1985.

Friedman, Alan J. "Contemporary American Physics Fiction." *American Journal of Physics* 47 (May 1979) 392–395.

Hoffman, Banesh. *The Strange Story of the Quantum*. New York: Dover, 1959.
An introductory, essentially nonmathematical discussion of quantum mechanics written for the general public.

Heisenberg, Werner. *Physics and Philosophy: The Revolution in Modern Science.* New York: Harper, 1958.
Presents the philosophical viewpoint of one of the great architects of the quantum theory.

Holton, Gerald and Stephen G. Brush. *Introduction to Concepts and Theories in Physical Science*, 2nd ed. Reading, Mass.: Addison-Wesley, 1965. Chaps. 26, 28, and 29.
See comment in references for Chapter 1.

Juki, Stanley L. *The Relevance of Physics*. Chicago: University of Chicago Press, 1966.
See comment in references for Chapter 1.

Lindsay, Robert Bruce and Henry Margenau. *Foundations of Physics*. New York: Wiley, 1936.
See comment in references for Chapter 6.

Chapter 8

Cohen, B. L. *Concepts of Nuclear Physics*. New York: McGraw-Hill, 1971.
An advanced undergraduate text on introductory nuclear physics.

Eisberg, R. and R. Resnick. *Quantum Physics*. New York: Wiley, 1974.
A less advanced undergraduate text with good introductions to the basic ideas of modern physics, including nuclear physics.

Enge, H. A. *Introduction to Nuclear Physics*. Reading, Mass.: Addison-Wesley, 1966.
Another well-written advanced undergraduate text.

Gamow, G. W. *Mr. Tompkins in Paperback*. New York: Cambridge University Press, 1971. Chaps. 12–15.
See comment in references for chapter 6.

Particles and Fields. San Francisco: Freeman, 1980.
A series of reprints from *Scientific American* on recent advances in the subject of elementary-particle physics, written for the interested general public. See also various articles in *Scientific American* written since 1980.

Index